Subantarctic Macquarie Island

Subantarctic Macquarie Island: environment and biology

P.M. SELKIRK,
R.D. SEPPELT
and
D.R. SELKIRK

The right of the
University of Cambridge
to print and sell
all manner of books
was granted by
Henry VIII in 1534.
The University has printed
and published continuously
since 1584.

CAMBRIDGE UNIVERSITY PRESS

Cambridge

New York   Port Chester

Melbourne   Sydney

CAMBRIDGE UNIVERSITY PRESS
Cambridge, New York, Melbourne, Madrid, Cape Town, Singapore, São Paulo

Cambridge University Press
The Edinburgh Building, Cambridge CB2 8RU, UK

Published in the United States of America by Cambridge University Press, New York

www.cambridge.org
Information on this title: www.cambridge.org/9780521266338

© Cambridge University Press 1990

First published 1990
This digitally printed version 2008

A catalogue record for this publication is available from the British Library

Library of Congress Cataloguing in Publication data

Selkirk, P. M.
Subantarctic Marquarie Island: environment and biology P. M.
Selkirk, R. D. Seppelt, and D. R. Selkirk.
    p.   cm. – (Studies in polar research)
Bibliography.
Includes index.
ISBN 0 521 26633 5
1. Natural history – Tasmania – Macquarie Island.   2. Macquarie
Island (Tas.)   I. Seppelt, R. D.   II. Selkirk, D. R.   III. Title.
IV. Series.
QH197.S37   1990
508.946–dc20   89-35796   CIP

ISBN 978-0-521-26633-8 hardback
ISBN 978-0-521-07603-6 paperback

Contents

Figures

Tables

Preface

Macquarie Island, a small subantarctic island, is politically part of the Australian state of Tasmania. From soon after its discovery in 1810 until 1920 its seal and bird populations were extensively exploited for pelts and oil. The events of this era have been documented in detail by Cumpston (1968). Now that the island is a Nature Reserve with UNESCO Biosphere status such exploitation is a thing of the past, and many aspects of the island's natural history are the subject of fascinated observation and scientific research. Much, of course, remains to be learned.

The account which follows is written by three biologists whose interest in and love for Macquarie Island has grown well beyond our initial studies of mosses there. We have drawn extensively on our own observations made during several summers' fieldwork on the island, as well as on the published work of others. While preparing this manuscript, we assembled an annotated bibliography of Macquarie Island (Selkirk, Selkirk & Seppelt, 1986).

We are grateful to the Macquarie Island Advisory Committee for permission to visit and conduct research on the island, and to the Australian Antarctic Division for logistic support during field trips. We are grateful to those at Macquarie University and at Antarctic Division who have provided direct assistance in preparing this manuscript: B. Duckworth and G. Rankin for drafting some of the diagrams; R. Oldfield, J. Norman, R. Reeves for photographic advice and processing; A. Gunawardena, the late P. Hughes, S. Hummelstad for typing and word processing assistance; J. Edgecombe, A. Downing, and F. Buining for many forms of general assistance. We are grateful to colleagues who have read and made valuable comments on parts of this book at the draft stage: D. Adamson, D. Bergstrom, M. Bryden, E. Colhoun, G. Copson, the late S. Greene, P. Greenslade, J. Joss, J. Peterson, P. Quilty, I. Stevenson, B.

Wellington. We record our thanks, also, to the many colleagues who, by their discussions over the years, have helped our ideas and our interest in the island to grow.

Finally, but most importantly, we express our thanks to our families who, by their encouragement and support, have made possible many months of fieldwork and many hours at the desk.

P. M. Selkirk

R. D. Seppelt

D. R. Selkirk

1

Introduction

Macquarie Island, a speck of land in the huge Southern Ocean, is a wild and beautiful place, windswept and lashed by stormy seas. It is of immense importance to the multitudes of seals and birds which, although they spend much of each year at sea, come ashore on the island to breed.

In addition to its intrinsic beauty, appealing to all who visit, the island is of immense interest to a variety of scientists. For geologists, it provides a unique example of seafloor material raised above sea level to become dry land in mid-ocean. The island has never been attached to any other land mass. To biologists, its plants and animals are of interest both as successful survivors of long-distance dispersal to the island and as tolerators of the harsh subantarctic climate. For meteorologists, climatologists and physicists, the island provides one of a very few land-based stations for monitoring southern high latitude climatic and atmospheric phenomena. It is of particular interest to ionospheric physicists since it is close to the southern auroral zone, and its geographic latitude is much less than its magnetic latitude.

Macquarie Island is one of six subantarctic islands or island groups: Macquarie Island, south of the Tasman Sea; South Georgia, south of the Atlantic Ocean; the Prince Edward Islands (Marion and Prince Edward), Iles Crozet, Iles Kerguelen, and the Macdonald Islands (Heard and Macdonald), south of the Indian Ocean (Figure 1.1).

All are oceanic islands, far from continental land masses. Their climates are strongly influenced by the Southern Ocean which surrounds them. All are close to the Antarctic convergence, an important oceanographic boundary where cold water from the ocean to the south meets warmer water from the north.

Macquarie Island, Marion Island, Prince Edward Island and Iles Crozet lie to the north of the Antarctic convergence; South Georgia, Heard Island and Macdonald Island lie to the south; Iles Kerguelen straddle it. All subantarctic islands experience cool, wet, windy conditions, but those to

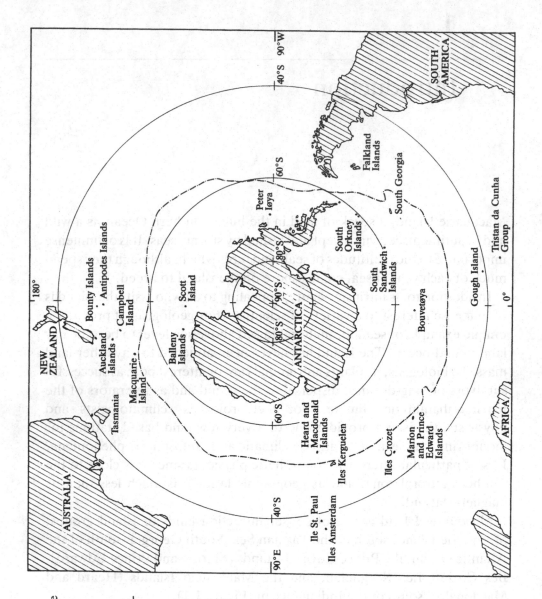

Figure 1.1. Antarctica and the Southern Ocean, showing location of subantarctic islands and Antarctic convergence. Based on the map of Antarctica and adjacent continents (Division of National Mapping, Australia, 1978).

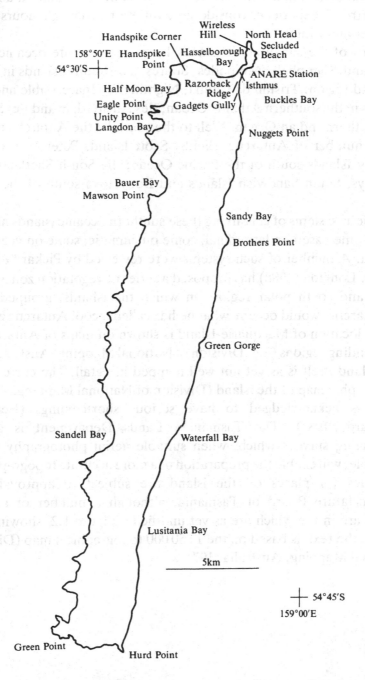

Wireless
Hill

Handspike Corner

North Head
Secluded
Beach

158°50'E Handspike
54°30'S Point

Hasselborough
Bay

ANARE Station

Half Moon Bay

Razorback
Ridge

Isthmus

Buckles Bay

Eagle Point

Gadgets Gully

Unity Point
Langdon Bay

Nuggets Point

Bauer Bay
Mawson Point

Sandy Bay

Brothers Point

Green Gorge

Sandell Bay

Waterfall Bay

5km

54°45'S
159°00'E

Lusitania Bay

Green Point

Hurd Point

Figure 1.2. Macquarie Island, showing places named in text. Based on map of Macquarie Island, Tasmania (Division of National Mapping, Australia, 1971). See also Figures 3.6, 7.3.

the south of the Antarctic convergence are somewhat cooler than those to the north. All experience considerable variation in daylight hours between summer and winter.

North of the subantarctic islands lie cool temperate oceanic islands; Auckland, Campbell, Antipodes, Snares and Bounty Islands in the New Zealand region; Tristan da Cunha, Nightingale, Inaccessible and Gough Islands in the southern Atlantic Ocean; Iles Amsterdam and Iles St Paul in the southern Indian Ocean. Well to the south lie the Antarctic continent and a number of Antarctic islands: Scott Island, Peter I Øya and the Balleny Islands south of the Pacific Ocean; the South Shetlands, South Orkneys, South Sandwich Islands and Bouvetøya south of the Atlantic Ocean.

Various systems of classifying these southern oceanic islands have been used, some based on latitudinal, some on climatic, some on vegetational criteria. A number of such systems were reviewed by Pickard & Seppelt (1984). Longton (1988) has proposed a series of vegetation zones for both north and south polar regions in which the islands, grouped here as subantarctic, would occupy what he has called a cool Antarctic zone.

The location of Macquarie Island is shown on maps of Antarctica and surrounding regions (e.g. Division of National Mapping, Australia, 1978). The island itself is as yet not well mapped in detail. The currently used topographic map of the island (Division of National Mapping, Australia, 1971) is acknowledged to have serious shortcomings (Berkery & Pritchard, 1987). The Tasmanian Lands Department is gradually conducting surveys which, when suitable aerial photography becomes available, will enable the preparation of a more accurate topographic map.

Names for places on the island are subject to approval by the Nomenclature Board of Tasmania, although a number of additional names are in use which are as yet unofficial. Figure 1.2, showing names used in the text, is based on the 1 : 50 000 topographical map (Division of National Mapping, Australia, 1971).

2

Discovery and human occupation

Macquarie Island, with its sea stacks, reefs and attendant islets to north and south, rises from the Southern Ocean about 1000 km south-east of Tasmania in the path of almost incessant strong westerly winds of the 'Furious Fifties' (Figure 1.1). Cold, wet and windy, a Nature Reserve under the control of the Tasmanian government, the island is a hauling-out and breeding refuge for southern elephant seals, a growing population of fur seals (Chapter 9) and hundreds of thousands of penguins and other sea-birds. The island's wildlife is protected. This was not always so. For over a century the island's shores were a slaughterhouse during a period when the mammals and birds of the Southern Ocean were pursued for furry skins, oil-yielding blubber fat, whalebone and penguin oil. Unlike some of the larger subantarctic islands such as South Georgia, Macquarie Island was never a base for whaling. Its coast offers no safe protected anchorages for land-based whaling ships and factories or even modern-day shipping (Figure 2.6).

Discovery and exploitation

Macquarie Island's recorded history began in 1810 when the *Perseverance*, a sealing ship out of Sydney in the colony of New South Wales, was blown off course during a voyage to sealing grounds on islands south of New Zealand; its crew chanced upon this previously unknown island whose shingle beaches were littered with fur seals and 'sea elephants' (southern elephant seals). A gang of men was put ashore to harvest fur seal pelts, while the *Perseverance* returned to Sydney for provisions and all-important supplies of salt for treating the pelts. A hundred years of commercial exploitation of Macquarie Island's marine mammals and penguins had begun.

The *Perseverance*'s crew may not have been the first humans to set foot on Macquarie Island. The *Sydney Gazette* of 5 January 1811 published an account of the island by Captain Smith of the American sealing ship *Aurora*:

> Captain Smith saw several pieces of wreck of a large vessel on this island, apparently very old and high up in the grass, probably the remains of the ship of the unfortunate de la Perouse.   (Cumpston, 1968: 15)

Mawson, too, referred to the wreck:

> Though the *Perseverance* was the first vessel to report the existence of Macquarie Island she was not the first to reach its shores, for the sealing gang left ashore by Hasselborough found portions of wreckage of a large vessel of ancient design, and apparently long cast up, high amongst the tussock-grass above the shore on the west coast. No clue was obtained as to the origin of the vessel, nor was there any indication of former occupation of the island.   (Mawson, 1943: 12)

Mawson's description of the 'ancient design' of the wreck on the island's west coast is interesting. We have been unable to find any reason for his decision that the vessel's design was 'ancient', nor what he actually meant by the term. The wreck's position could suggest a ship rounding the Cape of Good Hope only to be storm-driven onto Macquarie Island's west coast. It could represent either a hulk driven ashore or a shipwreck. If the latter, any survivors would have led miserable lives.

Sea surface currents and ocean drift in the Macquarie Island–New Zealand region are complex and the wreck may have arrived on the island's west coast from almost any direction. The wreck's origin will never be known as the relics which may have yielded clues to modern archaeologists subsequently disappeared – possibly burnt for firewood (Cumpston, 1968: 15).

The sealers of the *Perseverance* and other sealing ships (it being difficult to keep secret the location of islands with fur seal populations) were efficient. Fur seals were effectively wiped out on the island within a decade. So ruthless was the onslaught that even the type(s) of fur seal originally present on the island (Chapter 9) is now unknown.

With fur seals gone from Macquarie Island, at least as a commercial proposition, hunters turned to producing oil from blubber of southern elephant seals and, later, the fat of penguins. In a world where animal fats and oils lit houses, formed a base for paints and acted as lubricants, there was a steady and remunerative demand for seals and penguins rendered down to fit neatly into standardised barrels.

Eventually the production of oil from Macquarie Island's southern elephant seals and penguins ceased. Rusting try-pots, digesters and boilers now stand on the island's shores surrounded by distant descendants of some of the animal populations fed through them. The men and small ships involved in hunting for pelts and oil have gone, leaving behind names on the island map, a few graves, relics and shipwrecks (Figures 2.1, 2.2).

They also left behind additions to the island's original animal and plant life: mice, black rats, cats, wekas (Figure 2.3), rabbits and possibly three plant species. Wekas and rabbits were deliberately introduced. The island must be one of the few places in the world where black rats can be studied in the absence of competition from brown rats. The deliberate or accidental introduction of alien animal species led to extinction of two endemic birds: a flightless rail and a parakeet.

A very comprehensive history of Macquarie Island and the personalities involved in exploiting its animal populations from the time of discovery until its eventual declaration as a sanctuary in 1933 is given by Cumpston (1968).

Early scientific visitors

Macquarie Island's isolated geographical position meant that scientific interest in it developed slowly. Exploring expeditions from various nations called there while probing the icy outlines of Antarctica but results from their short visits were meagre. Thaddeus von Bellingshausen, commanding an expedition from imperial Russia, visited Macquarie Island in 1820. Naturalists who were to have accompanied his expedition had, unfortunately, never boarded his ships and he left the island after a brief stay, taking with him two albatrosses, twenty dead parakeets, one live one, and the skin of an elephant seal. Specimens of *Stilbocarpa polaris* (Macquarie Island 'cabbage' – Figure 2.4), used by sealers, ship's crews and early expeditions as an antiscorbutic, made their way back to St Petersburg. The United States Exploring Expedition of 1838–42 called at the island in 1840. Any specimens collected by the party sent ashore by expedition commander Wilkes were lost in the surf as the men returned to the ship. Wilkes commented: 'Macquarie Island affords no inducement for a visit' (Wilkes, 1845, quoted in Cumpston, 1968: 75).

Robert Falcon Scott, sailing from England to New Zealand on his way to Antarctica for the first time, landed on Macquarie Island with several companions and succeeded in shooting an elephant seal and specimens of several species of birds. In a hut on the island the party found a collection of prepared bird skins, including skins of the extinct endemic rail, but left the island without taking the collection with them (Cumpston, 1968: 209).

Before 1911 it was the individual traveller, prepared to brave wild seas

Figure 2.1. Oiling works and living huts at the Nuggets, 1913. E.R. Waite Album. Reproduced courtesy South Australian Museum.

Figure 2.2. Similar view as Figure 2.1: remnants of oiling works at the Nuggets, 1985. Moulting royal penguins around the beach. Since 1913 large areas of tussock have disappeared from the area.

in a tiny vessel, who contributed most to our knowledge of Macquarie Island. The first reasonably comprehensive report on the island was published in 1822 in Sydney, based on observations made by Thomas Raine, master of a vessel involved in sealing. Raine provided soundings at various locations around the island, commented on its geology and described plants, birds, seals and marine invertebrates.

Professor J.H. Scott of Dunedin, New Zealand, visited Macquarie Island in 1880 to make collections, and published a description of the island's flora and fauna (Scott, 1883). Another New Zealander, Professor A. Hamilton, also of Dunedin, worked on the island in 1894. Plant

Figure 2.3. Weka, introduced as a food supplement for sealers.

specimens collected by him made their way to the collections of the Royal Botanic Gardens, Kew.

To Hamilton belongs the rather dubious honour of being the first scientist to attempt deliberate modification of the island's flora.

> From the list given by Professor Scott, and the revision of the Macquarie Island plants published by Mr. T. Kirk . . . it was evident that there was very little chance of finding any useful shrubs; so before leaving Dunedin I determined to try and establish some on the island that might some day be of use, at any rate for firewood.   (Hamilton, 1895)

Hamilton took with him 'a large bag of seeds of several New Zealand *Pittosporums*, and of a variety of deciduous trees'. He also took seed of pines and 'a quantity of cabbage seed'. These various seeds he planted near Lusitania Bay and wrote 'I trust that some of them may become established'. Hamilton's experiment failed and there are still no trees, shrubs or cabbages on the island. Further experimental plant introductions have been tried without, however, the intention of modifying the native flora to provide a source of fuel and food for humans (Chapter 12).

The Australasian Antarctic Expedition (AAE)
Long-term scientific investigation of Macquarie Island began in 1911. The Australasian Antarctic Expedition 1911–14, led by Douglas

Figure 2.4. *Stilbocarpa polaris* – the Macquarie Island cabbage.

Mawson, established a major base on the island as a combined wireless relay station and meteorological observatory. Macquarie Island, more-or-less halfway between Tasmania and Antarctica, was an ideal position for such a base (Figure 1.1). The wireless station was set up at the northern end of the island, the only place where an easy landing can be effected, and the necessary masts and aerials were erected on what became known as Wireless Hill. This wireless/weather station functioned throughout the expedition, its staff (Figure 2.5) compiling weather reports and surveying and mapping the island which they shared with a few remaining animal oil producers (and killers of any fur seal they came across). L. R. Blake, surveyor and geologist, prepared a detailed topographic map of the island and established names for many of its features. Today's official map of the

Figure 2.5. The Macquarie Island party, AAE, 1911-14. Left to right: C.A.
Sandell, G.F. Ainsworth, A.J. Sawyer, H. Hamilton, L.R. Blake. Photograph
F. Hurley. Reproduced courtesy Mitchell Library, State Library of New South
Wales.

island bears names for some features which differ from those used by
Blake. Blake was killed at the close of the first world war and his
comprehensive field notes were not worked up for publication as a detailed
geographical and geological description of the island until 1943 (Mawson,
1943).

H. Hamilton began the first long-term study of the island's plants. His
collections formed the basis for a publication on the island's plants by
Cheeseman (1919). Ecological notes on the flora, accompanied by general
illustrations of the island's vegetation, were published by Hamilton
(1926).

When Mawson's expedition returned to Australia, the Macquarie
Island base established by him was taken over by the Australian
Commonwealth Meteorological Service which operated it until the end of
1915. When the base was closed, departing staff left in the care of the
headman of a small party still after animal oils, the remainders of the
ducks, fowls and sheep with which the meteorological station had been
equipped.

Control of sealing

The oiling industry on Macquarie Island had for many years been run from New Zealand. In 1889 the New Zealand government attempted to have Macquarie Island, apparently then unclaimed by any nation, formally annexed to New Zealand in an effort to control the activities of sealers and birders on New Zealand's southern shelf islands. Authorities in New Zealand realised that ships supposedly heading for Macquarie Island were just as likely to be poaching animals on other islands where such activity was forbidden, and recognised that policing of the whole area was necessary. The proposed annexation was approved by the Colonial Office in London, only to be aborted when it was belatedly realised that Macquarie Island was already a dependency of Tasmania, mentioned as such in the Letters Patent of the Governor of Tasmania. Negotiations between the governments of Tasmania and New Zealand to effect transfer of responsibility for the island were unsuccessful and Macquarie Island remains under Tasmanian jurisdiction. Tasmanian authorities had, however, recognised that control of exploitation of wildlife on the island was necessary and, in April 1891, Tasmanian Government Notice number 147, in accordance with the provisions of the Tasmanian Fisheries Act (1889), stipulated:

> 1. The taking of seals, whether known by the names seals, sea elephants, or sea lions, or any other local name upon Macquarie Island, and the islands adjacent there-to in the South Pacific Ocean, being dependencies of the Colony of Tasmania, is hereby prohibited.
> 2. Any person committing any breach of the aforesaid Regulation, or who shall buy or sell, or cause to be bought or sold, or have in his possession or control any seal so taken, shall be liable to a penalty of £5, and the forfeiture of all boats, engines or other instruments used in committing a breach of these Regulations.　　　　　　(Cumpston, 1968: 149)

Tasmania thus began exercising control over Macquarie Island which until then it had more or less ignored. The killing of wildlife, however, continued. In July 1891, the regulation promulgated in April of the same year was amended to read:

> The taking of Seals, whether known by the name of Seals or any other local name, in Tasmania and its Dependencies is hereby prohibited for a period of Three years . . .

The regulation as promulgated no longer applied to elephant seals, a source of oil, and nothing was said about penguins. Oiling gangs were thus free to return to the island.

Eventually, in 1902, an occupational licence, covering all Macquarie Island, was issued to New Zealand interests. In 1905 a seven-year lease of the island was granted to the same syndicate, terminating in 1911 because of non-payment of rent. The oiling gang at work on the island during the Australasian Antarctic Expedition's occupation of the wireless/weather base apparently had no legal right to be there. A further lease over the island, excluding the area occupied by the now-established weather station, was granted in 1915. The terms of the lease stipulated that king penguins were not to be molested, but allowed removal from the island of bird oil. Rent was £40 per annum, payable half-yearly in advance.

In 1915, Mawson and other members of the Australasian Antarctic Expedition began a campaign to have Macquarie Island declared a wildlife sanctuary and commercial exploitation of its wildlife stopped. Their work to achieve this aim started at a time when public attitudes to wildlife were changing and it was becoming more widely recognised that animals have values other than commercial ones.

The lessees and the Tasmanian government began to feel the effects of Mawson's campaign to have the island declared a sanctuary. Accusations of needless cruelty to animals slaughtered in the process of procuring oil were made by men who had visited the island, accusations hotly denied by those involved in obtaining the oil. An annual licence for removal of animal oil from the island was issued in 1918 and renewed for one year only in 1919, with the stipulation that all plant involved in oil-gathering be removed before termination of the allotted period or else forfeited. The Tasmanian government withstood pressure from oiling interests to allow the lease to continue and it was cancelled in 1920 (Cumpston, 1968).

Sanctuary

Mawson's wish that the island become a sanctuary for wildlife was realised in May 1933 when the Lieutenant-Governor of Tasmania signed a proclamation, under the provisions of the Tasmanian Animals and Birds Protection Act 1918, declaring Macquarie Island a wildlife sanctuary for birds and animals generally.

The conservation status of the island has undergone several changes since it was first declared a reserve. When the Tasmanian National Parks and Wildlife Service was established in 1971 the island became a conservation area, upgraded to a state reserve in 1972, extended to its present boundaries in 1978 and renamed Macquarie Island Nature Reserve. The reserve includes the entire island, the offshore rocks and attendant islets to north and south, to the low-tide mark. In 1977 Macquarie Island was declared a Biosphere Reserve under the UNESCO Man and the Biosphere Programme.

Part of the Tasmanian Shire of Esperance, Macquarie Island became a restricted area in 1981, under regulations of the Tasmanian National Parks and Wildlife Act (1970), controlled and managed by the Tasmanian Department of Parks, Wildlife and Heritage. Permission to land on the island must be sought from the Director of the Service by any expedition or person wishing to visit the reserve.

Regulations govern behaviour of individuals on the island. Collection of scientific specimens (whether plant, animal, rock or historic relic) requires a permit, as does immobilisation of any animal, possession or use of firearms or other hunting equipment. No visitor to Macquarie Island can deliberately introduce an exotic organism and all care must be taken to prevent accidental introductions. Visitors must take all precautions against disturbing or endangering any native plants, animals or their habitats.

ANARE – Australian National Antarctic Research Expeditions and Macquarie Island

Sir Douglas Mawson retained his interest in Antarctica and Macquarie Island throughout his life and often stressed that Australia should be seen to take an active interest in the Antarctic territories to which it laid claim. The British, Australian and New Zealand Antarctic Research Expedition 1929–31 (BANZARE) represented the only Australian activity in the region between the two world wars, a period when the Great Depression also made any plan to institute an active Australian programme in Antarctica and subantarctic regions difficult to realise.

After the second world war, Mawson achieved his long-term goal. Australian interest in lands to the south quickened; committees were formed and plans discussed. Stuart Campbell, a former member of BANZARE, was eventually appointed executive officer to organise an Australian expedition.

An Antarctic Planning Committee comprising Mawson, John King Davis (who had sailed with Shackleton in *Nimrod* and with Mawson in *Aurora*), naval and air force representatives and a member from the Council for Scientific and Industrial Research (now CSIRO) was established in 1947 to make recommendations on Antarctic and subantarctic activities to the Australian government. It was clear that government involvement was necessary. The days of Antarctic exploration funded largely from private sources, even if subsidised to some extent by governments, had disappeared.

The Australian government accepted a plan for a series of expeditions involving a number of activities. There was to be a systematic reconnaissance of the coast of the Australian Antarctic Territory over a number of years in order to choose a site for a scientific station on the Antarctic mainland. The real aim of any such base was political – an exercise of Australian sovereignty – but Mawson had always insisted that any such politically-motivated base should also serve a serious scientific purpose.

Out of the decision to mount the expedition grew the formalised bureaucratic institution of the Australian National Antarctic Research Expeditions (ANARE), part of an Australian government department, responsible for maintenance of Australia's Antarctic and subantarctic stations and the logistics of supplying and staffing them. ANARE has a small scientific staff of its own and regularly carries Australian and foreign scientists to its various stations.

Establishment of the Macquarie Island ANARE station

The Antarctic Planning Committee recommended that Australia gain access to a vessel specially fitted for Antarctic work (Campbell, 1949). Australia has always relied on chartering foreign-owned vessels for its Antarctic programmes. Many plans have been made either to build or buy an Australian Antarctic ship. The grounding on Macquarie Island and subsequent scuttling in December 1987 of MV *Nella Dan*, a Danish ship chartered for twenty-six years, may have precipitated the decision to let a contract for construction of an Australian-owned ice-breaking supply ship. *Nella Dan* was the most recent of a series of recorded shipwrecks on the island (Figure 2.6).

The Antarctic Planning Committee recommended coastal reconnaissance of the sector of Antarctica claimed by Australia with a view to selecting a site for a mainland base. It also recommended that scientific and weather stations be established on both Heard and Macquarie Islands, each to be operated for a period of at least five years. Heard Island was included in the plans at the instigation of Great Britain which then claimed sovereignty, later transferred to Australia. Reoccupation of Macquarie Island was suggested by Mawson.

Wyatt Earp – Lincoln Ellsworth's Antarctic veteran (then Australian-owned) which had begun life as a wooden North Sea herring trawler – was selected for the Antarctic reconnaissance voyage along the coast of Antarctica. Another vessel was needed to establish the two

Figure 2.6. MV *Nella Dan* aground in Buckles Bay, December 1987. Photograph P. Smart.

proposed subantarctic bases. The choice, despite severe criticism of her sea-worthiness under such conditions, eventually fell on HMALST *3501*, an LST (Landing Ship – Tanks) displacing about 3500 tons and designed to land tanks, other war material and men directly onto beaches in more placid seas than those of the subantarctic. HMALST *3501* sailed, creaking and groaning, to establish the bases – that on Heard Island first. Damaged during the Heard Island operation, the vessel returned to Melbourne for docking and repairs. On 3 March 1948, loaded with 400 tons of materials and supplies, men to establish the base and two army DUKWs (amphibious landing craft which could be swung over the side) to ferry the lot ashore, the ship sailed from Hobart to establish an Australian presence on Macquarie Island for the first time since 1915. The base established in 1948 has since been staffed continuously.

Macquarie Island's topography dictates where any base can be established, as found by Mawson in 1911 and by the sealers before him. The only comparatively safe anchorages with relatively easy access to an area of flat ground are at the extreme northern end of the island. There, a narrow low-lying isthmus, across a low, narrow section of which waves wash during the heaviest seas, joins the main mass of Macquarie Island to the isolated flat-topped hill on which Mawson and his party had set up their wireless masts in 1911. Mawson's huts had been tucked away under the comparative shelter of the base of Wireless Hill where the flat isthmus gives way to the steep slopes of the hill. The new station was established in the same place in 1948.

When the LST had discharged the last load and left the island, thirteen men were left in splendid physical isolation to carry out research and man radio and weather stations throughout the following winter and summer. The work schedule was a busy one. Weather records were kept, balloons carrying a radiosonde sent aloft daily and synoptic reports transmitted to Australia.

In planning the first Australian National Antarctic Research Expedition there had been much discussion of priorities to be given to various types of research. It was decided there was no urgent need to undertake extensive biological work, felt to have been adequately covered by the Discovery Committee, organiser of BANZARE. Research during the initial period of establishment and maintenance of the Macquarie Island station favoured the physical sciences. Active programmes in meteorology, geo-magnetism, seismology, cosmic ray research, auroral physics and radio-physics, begun when the base was established, have continued.

In July 1948 tragedy struck the party when the diesel engineer, C. H. Scoble, skiing on the plateau, drowned in the lake which today bears his name. Arrangements were made for another engineer to join the island party, the new engineer travelling not by ship but by aeroplane – a Catalina.

At that time the Royal Australian Air Force was making trials of JATO (Jet Assisted Take-Off) on Catalinas. Catalina A24-104, based at Rathmines on Lake Macquarie just south of Newcastle, New South Wales, was equipped with four JATO units and flew to Hobart where she waited thirteen days for suitable weather to fly to Macquarie Island. Two attempts to reach the island were aborted when the weather deteriorated. The eventually successful flight to Macquarie took seven and a half hours and Scoble's successor was put ashore. The JATO apparatus was used to get A24-104 airborne from choppy waters and the aircraft returned to its base at Rathmines via New Zealand, a deviation made necessary because of prevailing westerly winds.

It is ironic that Macquarie Island, now that long-range amphibious aircraft have disappeared from the skies, should be more isolated in 1988 than in 1948. No runway able to handle modern aircraft capable of the return flight to Macquarie from either the Australian mainland or Tasmania can be built on the island. The Isthmus, the only reasonably flat area, is short and ends in steep bluffs at each end. Building an airstrip on the plateau with its rugged topography is impossible. The only possible access today is by ship. The Australian Air Force makes regular air-drops (begun in 1978) but have never off-loaded a passenger by that technique. The only other aircraft to grace the skies over Macquarie are helicopters carried on Antarctic supply ships or Australian naval ships. None is based on the island.

The present ANARE station

Various observatories on Macquarie Island are directed by different Australian government agencies. The geophysical observatory is the responsibility of the Bureau of Mineral Resources, Geology and Geophysics. Staff of the Australian Bureau of Meteorology carry out meteorological tasks. The Antarctic Division conducts its own upper atmospheric physics research programme as well as running programmes for the Ionospheric Prediction Service.

Research programmes in biology, geology and geomorphology, not supported by particular government agencies, were added to the list of activities on Macquarie Island a few years after 1948, and the wealth of scientific data now available forms the basis for the chapters which follow.

The ANARE station, accommodating eighteen to twenty people during winter and coping with extras in summer and during a ship's visit, covers the site of Mawson's 1911 base. Prefabricated buildings, hastily erected to house the 1948 party, have been replaced by a straggling collection of buildings for various purposes (Figure 2.7).

Three glasshouses grow herbs and vegetables for domestic purposes. The only road on the island is a short track along the Isthmus. Despite the fact that the station is on an island, no boat is kept there, a reflection of the dangerous nature of the shoreline and surrounding seas. Six field huts, equipped with basic facilities of bedding, food, heat, light, emergency medical supplies and radio (Figure 2.8) are scattered around the island, the furthest about ten hours' walk from the main base along a system of foot tracks.

End of an era in the far south

Two very brief, bald statements in *Polar Record* point to the beginning of a new era in Australian Antarctic and subantarctic work.

Figure 2.7. The ANARE station at the northern end of the Isthmus, 1984. Note the southern elephant seal wallow pools amongst tussocks on the west (left of figure) side of Razorback Ridge in the foreground. The flat top of Wireless Hill (in the background) resulted from a period of marine planation when it was close to sea-level before rising to its present altitude.

ANARE relief voyages and personnel for the summer of 1959-60 are listed. Under the heading 'Macquarie Island ' all male personnel of the expedition are named. The list of names is followed by the statement that:

> Four women biologists carried out work in littoral ecology, the effects of nesting sea-birds on plants, and the ANARE bird and seal programme.
>
> (Law, 1961: 398)

The 'women biologists' remain anonymous.

The *Polar Record* report of ANARE relief voyages of 1960-61 refers to Macquarie Island and to the fact that:

Figure 2.8. Caroline Cove field hut at the foot of Mt Haswell, 1984.

> Australian women biologists made botanical collections and investigated
> the littoral ecology of the island, . . . (Law, 1962: 185)

Female anonymity is again preserved! The 'anonymous females' have,
however, had their say.

> My own modest claim to fame is that, while Biological Secretary to the
> Antarctic Division . . . I walked into Phil Law's [P. G. Law, former
> Director of ANARE] office one day in 1959, asked to go to Macquarie,
> and caught him at exactly the right moment. We easily made up a cabin
> of 4 women for the December 1959 changeover. Isobel Bennett and
> Hope Mackenzie (littoral ecologists) and Mary Gillham (cryptogamic
> botanist from England) and me . . . After that women went on nearly
> all changeovers to Macquarie (a toe in the Antarctic door, if not a foot).
> (Ingham, 1984)

Between them, these women, the first to visit Macquarie Island with an official Australian expedition, produced two books on general aspects of the island and a series of scientific papers on various aspects of island biology.

Including women in an official expedition seems to have caused some trepidation amongst bureaucrats and the pioneering women themselves. A few male expeditioners to the subantarctic and Antarctica still maintain that these areas are no place for women, no matter how competent.

Isobel Bennett, one of Australia's foremost marine scientists, wrote:

> On that initial trip, it behoved the four women to tread warily. We were invaders in a man's realm and were regarded with some suspicion. We had been warned that on our behaviour rested the future of our sex with regard to ANARE voyages, an attitude which did not particularly amuse us. The venture inevitably made headlines. Today our presence on a polar ship south-bound for Macquarie does not even warrant a press mention, and we have become an accepted part of the scheme of things . . .
>
> (Bennett, 1971)

The initial party of four women to visit Macquarie Island must have shown exemplary behaviour. Women are now members of almost all Australian subantarctic and Antarctic expeditions. Initially women were only summer visitors to subantarctic and Antarctic latitudes. They now regularly overwinter with Australian expeditions. In 1976, Zoe N. C. Gardner was the first woman to overwinter with an Australian expedition, as medical practitioner on Macquarie Island.

3

The island and its setting

Physiography

Macquarie Island is elongate and almost rectangular, 34 km long by up to 5.5 km wide. Its long axis trends N15°E along the general line of the submarine ridge of which it and its outlying islets are uplifted fragments. Judge and Clerk Islands lie 14 km to the north, Bishop and Clerk Islands 33 km to the south. These islets, mostly barren rock less than 50 metres high, are geologically similar to the main island. They are poorly known because of difficulty of access (Lugg, Johnstone & Griffin, 1978).

The main island consists essentially of an undulating plateau more than 200 metres above sea-level (asl) (Figure 3.1). The plateau surface is dotted with lakes, tarns and streams, and its margin is bounded by scarps which fall abruptly to a narrow, low-lying coastal fringe. Plateau scarps are steep, generally at angles of 40° to 45°, but up to 80° in places (Figure 3.2). The west coast is more rugged than the east, indented with small bays and coves and fringed with lines of residual sea stacks and reefs.

A coastal terrace, formed from an old wave-cut platform now raised above sea-level, occurs on some parts of the island. It is best developed in the north-western corner of the island at Handspike Point where it is over a kilometre wide and up to 15 metres above sea-level (Figure 3.3). Relict sea stacks occur on this terrace, and reefs and sea stacks extend seaward from it, continuing the line of plateau spurs. The terrace continues for almost 10 kilometres southwards along the west coast from Handspike Point but then disappears, reappearing in only a few isolated locations further south. On the east coast a coastal terrace occurs intermittently in narrow strips. An area of terrace at Hurd Point, the south-eastern tip of the island, is occupied by a huge penguin colony.

A low-lying isthmus, about 200 metres wide and wave-swept in parts during severe storms, joins the otherwise isolated, flat-topped mass of Wireless Hill to the main body of the island (Figure 2.7).

The general surface of the plateau is interrupted by two narrow strips of

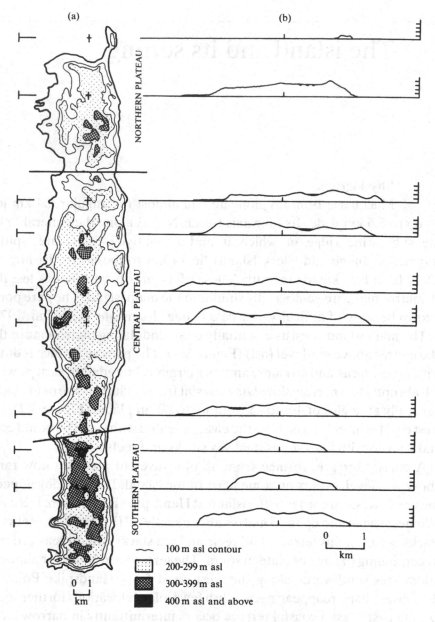

Figure 3.1.(a) Macquarie Island, contour interval 100 m; (b) east–west profiles through positions marked + on contour map. Horizontal scale shown, vertical scale: 1 unit = 100 m. Lakes shown in black in (b).

Figure 3.2. The rugged south coast of the island. Steep slopes below the plateau margin support tall tussock grassland. Lake Ainsworth lies in a fault-bounded depression. Two photographs form this mosaic (junction arrowed). Antarctic Division photograph, S. Brown.

Figure 3.3. A raised coastal terrace fringes cliffs which bound the plateau. Oblique aerial photograph of Macquarie Island from the north-west taken in 1947. This photograph is Crown Copyright and has been reproduced with the permission of the General Manager, Australian Surveying and Land Information Group, Department of Administrative Services, Canberra, ACT, Australia. Photograph ANT65:0038.

land, less than 200 metres in altitude, running across the island (Figure 3.1). One such area of lower land extends from Bauer Bay to Sandy Bay and is largely occupied by the valleys of Stony and Finch Creeks. The other low-lying strip lies south of Mt Tulloch, linking the valley of Flat Creek to the basins of Tulloch and Prion Lakes and the area around Brothers Lake on the east coast. The plateau area around Mt Tulloch is thus isolated from the main plateau. For purposes of description, the plateau can be divided into three main sections: northern plateau, central plateau and southern plateau (Figure 3.1).

The northern plateau is generally 200 to 250 metres asl with some areas exceeding 300 metres. The central plateau, also generally 200 to 250 metres asl with isolated areas over 300 metres, has two large basins in its flanks. Skua Lake occupies the western basin; Green Gorge the eastern (Figures 3.1, 3.4). A number of the island's larger lakes, Gratitude, Flynn, Major and Tiobunga Lakes, occur on the central plateau: all between 200 m and 300 m asl (Figures 3.5, 7.3).

The southern plateau is the highest part of the island, with a general surface more than 300 m asl. Three peaks reach more than 400 m; Mt Hamilton, the highest point on the island, reaches 433 m (Figure 3.6). Another large lake, Waterfall Lake, occurs on the southern plateau (Figure 7.3).

The very large expanse of ocean surrounding Macquarie Island has affected and continues to affect the island in many ways: geologically the island is of sea-floor origin, an above-sea portion of the Macquarie Ridge; climatically it is hyperoceanic; biologically it is an oceanic island which has never been connected to an adjacent land mass and whose organisms must have reached it by long-distance dispersal (Bergstrom & Selkirk, 1987). Understanding the oceanic environment in which it is situated is necessary to understanding the island itself.

Southern atmosphere

At 54°S, Macquarie Island lies within a zone known, since the days of sailing ships, as the 'Furious Fifties', because of its strong winds and stormy seas. The well-defined westerly wind circulation between about 35°S and 60°S has an average near-surface windflow of 6 to 7 m s^{-1} which varies little throughout the year (Phillpot, 1964).

A belt of low atmospheric pressure surrounds Antarctica at 62°S to 65°S. This low-pressure belt is the place of origin of numerous cyclonic depressions. These latitudes south of the Indian Ocean and of Australia constitute a major region of cyclone formation. East of this, there is a major region of cyclone decay which includes Macquarie Island (Kep, 1984). Other cyclones originate over warm ocean currents in middle latitudes, move south-eastwards and intensify. As Schwerdtfeger (1984) writes, 'cyclonic disturbances . . . can produce over the southern ocean surface winds and a state of the sea surpassing what can be found on other oceans'.

Macquarie Island, in the path of these disturbances (Figure 3.7), regularly experiences high winds. In the year from October 1983 to September 1984, strong winds (windspeed between 11.3 and 17 m s^{-1}) were recorded on 273 days, and gales (windspeed greater than 17.5 m s^{-1}) on 76 days. Winds come overwhelmingly from the north-west and west throughout the year. Mean annual windspeed recorded at Macquarie Island is 9.3 m s^{-1} (Table 3.1).

Figure 3.4. Green Gorge basin from the north. Sawyer Creek flows from the south across Green Gorge basin to the sea via a notch in the coastal ridge. Stepped benches are clear in coastal ridge to south. Composite of two photographs.

Figure 3.5. Major Lake, the largest lake on the island, lies between Mt Martin and the western edge of the plateau. The slopes in the foreground fall steeply to the western coast. Australian Antarctic Division photograph, S. Brown.

Southern Ocean

Within the Southern Ocean, circulation patterns and boundaries of different water masses are discernible on the basis of water temperature and salinity.

When seawater freezes, the salt is excluded, so that sea-ice is mostly frozen freshwater with some salt-water trapped in pockets. The overall salinity of sea-ice is about 6 parts per thousand and that of seawater in the region is between 34 and 35 parts per thousand. As sea-ice forms, there is an increase in salinity of the seawater from which it is freezing out. As sea-ice melts, there is a decrease in salinity of the seawater to which fresh melt-water is being added.

Variations in salinity and temperature cause formation of a number of distinct water bodies of differing densities in the Southern Ocean (Figure 3.8). They remain fairly distinct, despite some mixing at their boundaries. Southern Ocean water circulates horizontally around the globe as currents, while vertically it forms convergences and divergences (Figure 3.9).

The direction of current flow generally matches that of prevailing surface winds. Close to the Antarctic coast, the Antarctic coastal current

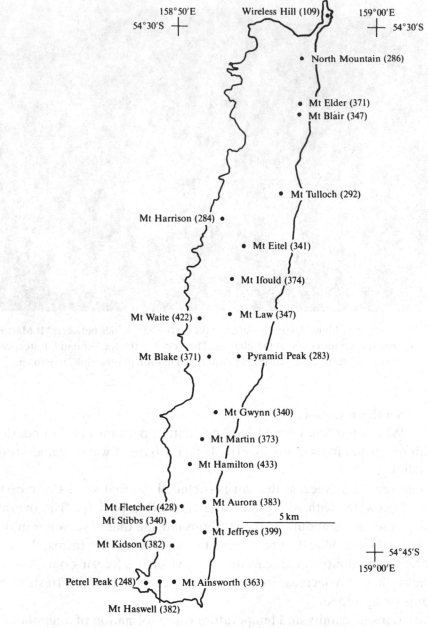

Figure 3.6. Named peaks on Macquarie Island. Altitude in metres above sea-level.

Figure 3.7. This infrared cloud picture from NOAA 9 satellite, transmitted from Washington to the Bureau of Meteorology, Melbourne, shows cloud associated with cyclonic depression in Macquarie Island region. Reproduced courtesy Bureau of Meteorology, Australia.

flows west, while north of about 60°S the Antarctic circumpolar current flows east. Numerous gyres occur between these two major currents. Sea-floor topography affects details of current direction in certain areas (Deacon, 1966). Macquarie Island lies within the region of the east-flowing Antarctic circumpolar current.

Five water bodies are of interest in considering vertical circulation between sea-floor and sea-surface in the Southern Ocean (Figures 3.8, 3.9). Antarctic bottom water (cold, saline and dense) originates in the Weddell and Ross Seas. Antarctic circumpolar deep water, highly saline but warmer and hence less dense, wells up at the Antarctic divergence and

Table 3.1. *Climatic summary for Macquarie Island, 1948–86.*

	Jan.	Feb.	Mar.	Apr.	May	June	July	Aug.	Sep.	Oct.	Nov.	Dec.	Year
Temperature (°C) mean[a]	7.0	6.9	6.3	5.2	4.2	3.3	3.3	3.3	3.4	3.9	4.6	6.1	4.8
extreme max.[b]	11.7	12.4	10.8	9.4	9.7	8.7	8.3	7.8	8.6	10.3	10.7	12.1	
mean max.[b]	8.4	8.3	7.6	6.6	5.6	4.9	4.7	4.9	5.2	5.5	6.3	7.7	6.3
extreme min.[b]	0.6	−0.1	−0.9	−4.5	−6.8	−7.0	−8.9	−8.9	−8.7	−4.6	−3.3	−1.4	
mean min.[b]	5.1	4.9	4.4	3.2	2.4	1.4	1.4	1.5	1.3	1.8	2.6	4.2	2.9
Precipitation (mm) mean total[b]	83	77	92	87	76	73	66	62	70	70	67	72	895
Evaporation (mm) mean[c]	45	38	31	31	27	30	29	28	33	56	46	49	445
Mean relative humidity (%)[b]	88	90	91	90	89	90	92	92	88	85	86	87	89
Mean daily sunshine (h)[b]	3.3	3.3	2.7	1.8	1.4	0.8	0.9	1.4	2.1	2.6	2.9	3.2	2.2
Radiation (kv cm⁻²) total[c]	47.8	36.3	25.8	13.9	6.5	3.3	5.3	9.5	21.1	32.5	41.5	49.4	292.9
Mean cloud cover (octas)[b]	7	7	7	7	7	7	7	7	7	7	7	7	7
Windspeed (m s⁻¹) mean[b]	8.8	9.4	9.9	9.9	9.4	9.4	8.8	9.9	10.4	9.9	8.3	7.8	9.3
max. gust[b]	32.9	32.8	35.2	35.7	35.2	36.7	33.3	34.6	35.7	35.6	30.6	30.1	

[a] From Jacka, Christou & Cook (1984) and records kept on Macquarie Island (data 1948–86 inclusive).
[b] From Seppelt (1980a) (data 1949–77 inclusive).
[c] From Jenkin (1975) (data 1967–69).

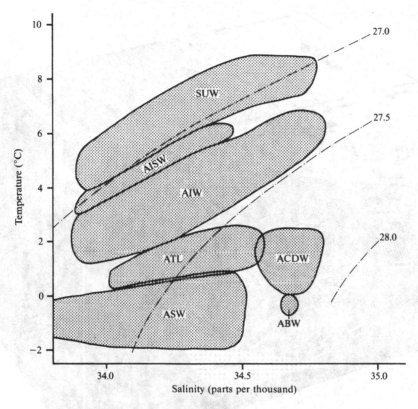

Figure 3.8. Temperature–salinity characteristics for Southern Ocean regional water masses. ᐧᐧᐧᐧᐧ — σt (measure of density). SUW = subantarctic upper water, AISW = Antarctic intermediate surface water, AIW = Antarctic intermediate water, ATL = Antarctic transition layer, ASW = Antarctic surface water, ACDW = Antarctic circumpolar deep water, ABW = Antarctic bottom water. Redrawn from Houtman (1967) with permission from the New Zealand Institute of Oceanography.

circulates north and south between Antarctic bottom water and surface waters. Close to Antarctica, Antarctic surface water varies seasonally in salinity because of freezing and melting of sea-ice. At the Antarctic convergence, south-flowing subantarctic surface water meets north-flowing Antarctic circumpolar deep water (Figure 3.9). As Antarctic circumpolar deep water sinks below the less-dense subantarctic surface water, some mixing occurs, forming Antarctic intermediate water (Foster, 1984).

The Antarctic convergence is an important oceanographic feature. It is recognised at the surface by a change from about 4 °C to 2 °C in sea-surface temperature over the relatively short distance of about 150 nautical miles.

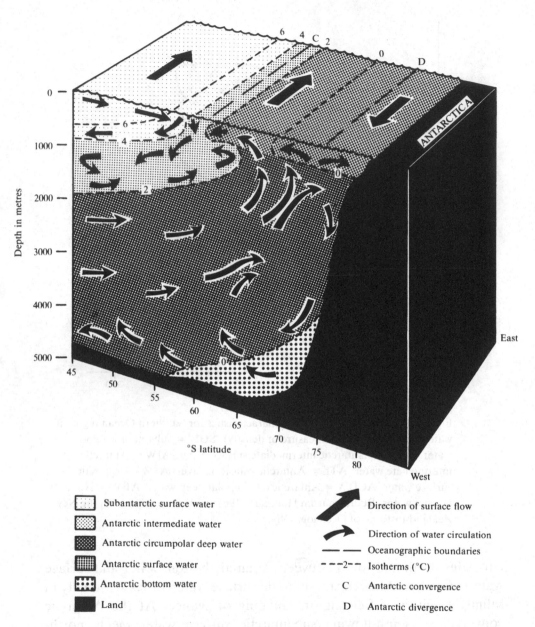

Figure 3.9. Schematic representation of Southern Ocean water bodies, currents and the Antarctic convergence. Drawn using information from Deacon (1966), Gordon & Goldberg (1970), Houtman (1967).

It can be recognised below the surface as the position at which the minimum temperature layer begins a rapid descent or falls below 200 m depth or the position at which minimum salinity at a depth of 200 m is found. Surface and subsurface expressions of the convergence usually occur at the same latitude (Deacon, 1966; Gordon, 1967; Gordon & Goldberg, 1970).

The mean position of the Antarctic convergence varies, moving within a zone of 2° to 4° of latitude. The actual position at any time lies within 60 to 120 nautical miles of the mean position usually shown on maps (Figure 1.1). The convergence is an oceanographic boundary determined by the circulation of water bodies. However, its position within its usual zone of occurrence may be affected by atmospheric conditions. When the circumpolar low-pressure zone is strongly developed and strong westerly winds promote northerly movement of Antarctic surface water and hinder southward movement of subantarctic surface water, the convergence occupies a northerly position. When the circumpolar low-pressure zone is weakly developed, the convergence moves southward (Buynitskiy, 1974).

Two maps published by the New Zealand Oceanographic Institute show the variation in position of the Antarctic convergence in the vicinity of Macquarie Island. On a map of currents south of New Zealand based on data collected in December 1956 to January 1957, Burling (1960) shows the Antarctic convergence just to the south of Macquarie Island. On a map showing temperature and salinity distribution in the same region based on data collected in November 1958, Houtman (1965) plots the Antarctic convergence just to the north of the island.

The Antarctic convergence is a significant biological boundary in the Southern Ocean. Many planktonic organisms, fish and birds are typical of either the southern or the northern side of the convergence and are rare on the other side. Not only do salinity and temperature conditions differ on each side of the convergence, but water circulation patterns maintain a high degree of isolation (Deacon, 1966).

The Antarctic convergence does not occur at the same latitude at all longitudes. It is about 15° further north in the east Atlantic Ocean than in the east Pacific Ocean (Figure 1.1). Large volumes of Antarctic bottom water flow north from their source in the Weddell Sea into the Atlantic sector of the Southern Ocean. The convergence generally tends to occur further to the north over submarine ridges and further to the south over deep basins, reflecting variations in position of bottom currents (Deacon, 1966).

South of Australia, the Antarctic circumpolar current flows eastward near 50°S. In the region of Macquarie Island, the north–south trending Macquarie Ridge, and the Campbell Plateau extending to New Zealand, form a shallow barrier to passage of the Antarctic circumpolar current. The current is deflected southwards, and passes through passages in the ridge at 53°30'S, 56°S and 60°S (Gordon, 1972). Southwards deflection of the Antarctic circumpolar current carries water characteristics some 10°S further south than they are usually found. Water circulation in upper layers of the ocean, and hence the position of the Antarctic convergence, are affected.

The position of the six subantarctic island groups relative to the Antarctic convergence is given in Figure 1.1.

Climate

The climate of Macquarie Island is cool, moist and windy and, because of the moderating influence of the extensive surrounding ocean, remarkably uniform. The island is said to have one of the most equable climates on Earth, although not one that humans commonly regard as pleasant.

A thermoisopleth diagram (Figure 3.10) makes clear the small range of temperatures the island experiences throughout a day (vertical axis) and throughout a year (horizontal axis). Wind, cloud cover, precipitation and relative humidity similarly vary very little throughout the year (Table 3.1). Precipitation occurs as mist, rain, sleet, hail or snow at any time of the year, often in several forms on the same day. Snow rarely lies on the ground for more than a week. The only marked seasonal variation is in daylength, which ranges from seven hours in July (winter) to seventeen hours in January (summer) (Figure 3.10).

Although light levels are low and hours of sunshine few, the island's plants appear able to utilise the available light efficiently at prevailing temperatures. Temperatures, though low, are not extreme enough to limit metabolic activity, and plant productivity is high (Jenkin & Ashton, 1970; Jenkin, 1975).

Global surface air temperatures over both land and sea have risen by about 0.5 C° during the twentieth century (Folland, Parker & Kates, 1984; Jones, Wigley & Wright, 1986). Temperatures at high latitudes in the northern hemisphere are amplified compared with those at lower latitudes (Kelly *et al.*, 1982; Raper *et al.*, 1983). In the southern hemisphere, Jacka, Christou & Cook (1984) noted a warming trend which was more pronounced at higher latitudes.

Meteorological data are sparse from many parts of the Southern Ocean,

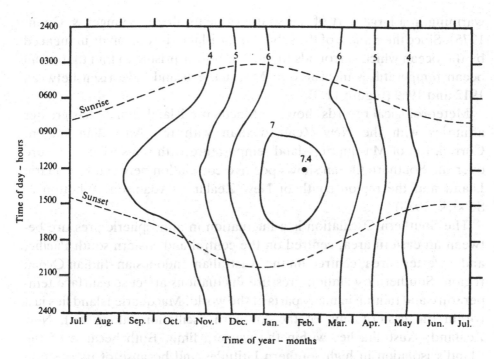

Figure 3.10. Thermoisopleth for Macquarie Island, 1957–68. Temperature variation throughout day, vertical axis; temperature variation throughout year, horizontal axis; 4, 5, 6, ... = °C . --- = sunrise, sunset. Redrawn using information from N. Wace (pers. comm.), Jenkin (1975).

particularly from south of 45°S. In this context, Macquarie Island data are of particular interest. Mean annual surface temperature has risen by one degree Celsius (1°C) since the start of temperature records in 1949. Comparison with trends at southern stations near the longitude of Macquarie Island confirms a warming trend progressively more pronounced at progressively higher latitudes (Adamson, Whetton & Selkirk, 1988).

Warming of Macquarie Island has been occurring in all months, but is most pronounced (0.042 C° yr^{-1}) in late summer and autumn and least pronounced (0.018 C° yr^{-1}) in spring. In other words, the rate of warming is greatest in those months when the sea-ice is furthest away (Jacka, 1983) and the subtropical high-pressure zone is closest (Pittock, 1971), presumably because of the greater likelihood of warm northern air reaching the island.

Temperature records from other Southern Ocean islands (Jacka, Christou & Cook, 1984) and the major retreat of Heard Island glaciers since the 1950s (Allison & Keage, 1986) support the suggestion of recent warming in a large part of the southern hemisphere (Salinger & Gunn, 1975). Since the climate of the subantarctic islands is so strongly influenced by the ocean which surrounds them, it is not surprising to find that mean ocean temperatures measured at Macquarie Island have risen between 1912 and 1969 (Figure 11.3).

Meteorological records show the Macquarie Island climate has greater affinities with the New Zealand than with the Australian region. Correlation of Macquarie Island temperature with sea-surface pressure over the Southern Ocean shows positive correlation between Macquarie Island and the region south of New Zealand (Adamson, Whetton & Selkirk, 1988).

The Southern Oscillation is a fluctuation in atmospheric pressure between an eastern area, centred on the central and eastern south Pacific, and a western area, centred on the Australian–Indonesian–Indian Ocean region. Southern oscillation pressure fluctuations affect sea-surface temperature and rainfall in many parts of the world. Macquarie Island lies just within the eastern limb of the southern oscillation, along with New Zealand; Australia lies within the western limb. Both because of the island's isolation in high southern latitudes and because of its position close to this major boundary in global atmospheric circulation, the records from the Macquarie Island meteorological station are of enormous interest in the regional and global sense (Adamson, Whetton & Selkirk, 1988).

Table 3.1 summarises data collected for the Australian Bureau of Meteorology on Macquarie Island since 1948. The meteorological observatory is 6 m asl on the Isthmus at the northern end of the island (Figure 2.7). Although these data tell us much about the Macquarie Island climate in the regional sense, since most of the island is a plateau more than 200 m asl, additional data recorded on the plateau are particularly useful in providing information about conditions on the island itself.

Although there has been no long-term systematic recording of climatic data at plateau sites, records have been kept for brief periods on several occasions. Ainsworth recorded temperatures during visits to various sites around the island in 1912 (Ainsworth, Power & Tulloch, 1929). If these are compared with temperatures recorded with simultaneous records at the meteorological observatory on the Isthmus, a lapse rate (drop in temperature with increase in altitude) of approximately 1 C° for each 100 metres altitude can be calculated (Taylor, 1955a). Jenkin (1975) also reported a decrease in air temperature of approximately 1 C° per 100 m increase in altitude. Data (Table 3.2) collected during the 1980–81 summer (P. Bannister, pers. comm.) allows comparison of maximum and minimum

Table 3.2. *Meteorological data, November 1980 to January 1981, from observatory on Isthmus and northern end of plateau.*

	Isthmus	Plateau site[a]	Comparison of plateau and isthmus records
November 1980			
Total precipitation (mm)	75.0	100.9	+35%
Maximum temperature (°C)	8.0	6.4	
Minimum temperature (°C)	−0.3	−2.0	
December 1980			
Total precipitation (mm)	34.4	47.9	+39%
Maximum temperature (°C)	10.0	7.9	
Minimum temperature (°C)	0.3	−0.5	
January 1981			
Total precipitation (mm)	129.6	166.2	+28%
Maximum temperature (°C)	10.7	8.4	
Minimum temperature (°C)	3.1	−2.0	

[a] Data from P. Bannister (pers. comm.).

monthly temperatures at 200 m asl with those at 6 m asl. The temperature drop approximates 2 C° for a 200 m increase in altitude.

Precipitation on the plateau is higher than at the meteorological observatory. Jenkin (1975) reported 16% more precipitation at the northern end of the plateau at 235 m asl. Bannister (pers. comm., Table 3.2) recorded 30% to 40% more precipitation at the same site in a different year; Mallis (1988) reported 42% more precipitation at the same northern plateau site and 25% more at 100 m asl on Wireless Hill than at the Isthmus observatory.

Windspeeds, too, are greater on the plateau than at the Isthmus observatory. Taylor (1955a, b) estimated them to be 25% greater and possibly much more, Jenkin (1975) 32% greater and Löffler (1983) up to 100% greater.

Using a flag tatter technique, Peterson & Scott (1988) examined wind run along a cross-island transect. Although they were not able to calibrate their flag tatter data in terms of windspeed or total windrun, it is clear that wind effects are very variable across the plateau, and considerably greater than close to sea-level on either the east or west coast or on the Isthmus.

Despite the few measurements reported, it is clear that plateau sites on Macquarie Island experience a climate significantly more severe than that

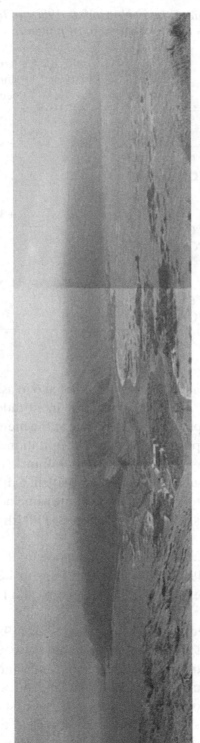

Figure 3.11. A cap of lenticular cloud frequently shrouds Macquarie Island's plateau, contributing to greater precipitation at plateau than at sea-level sites. Composite of four photographs.

documented by the Bureau of Meteorology at its official observatory. Plateau soil and air temperatures are lower, windspeeds and precipitation higher (Figure 3.11). These factors are of considerable importance to geomorphic processes and vegetation distribution on the island.

4

The island's origin and geology

In origin, Macquarie Island is unique among subantarctic islands. Its origins owe nothing to large-scale sedimentation as do South Georgia and the Kerguelen Archipelago. Nor is it an island on which there has been subaerial volcanic activity leading to further island-building as on the Prince Edward Islands, Heard and Macdonald Islands and Iles Crozet.

Oceanic in both climate and position, Macquarie Island is also totally oceanic in origin. The only geological deposits on the island which formed subaerially are very recent ones:

(a) beach deposits,
(b) a mantle of rock debris, peat and vegetation which obscures much of the detailed geology of the island, and
(c) lacustrine deposits.

Sections through old lake beds are exposed in the coastal scarps (Chapter 5). New sedimentary sequences are presumably now being deposited in the island's numerous lakes. Pieces of sandstone and granite found on beaches by members of Mawson's expedition are foreign to the island and have nothing to do with its origin or geology. They represent either material dumped by stranded icebergs (although sightings of icebergs from the island are few) or brought to the island in ships as ballast or for some other purpose (Cumpston, 1968).

Seismicity

Earthquakes and tremors have been noticed by visitors to the island since its discovery. These provide clues to the island's origin and help shape its present surface.

The leader of a sealing party in 1815 wrote:

> The first which took place on 31st October, 1815, at one o'clock in the afternoon, overthrew rocks, and gave to the ground the motion of a wave

for several seconds. Several men were thrown off their legs, and one was considerably hurt by his fall, but soon recovered. At two o'clock the same afternoon another earthquake was felt, another at four o'clock and ten shocks during the night, all of which were accompanied with a noise in the earth like that of distant thunder. On the 1st of November another shock was felt; and as the people were employed in distant divisions, their observations of the effects produced by the phenomena were most general. An overseer of a gang states that he witnessed the falling of several mountains, and the rocking of others which appeared to have separated from the summit to the base. On the 3rd of November, hard frost and heavy snow, two very severe shocks were felt. The 5th, 9th and 11th were attended with some alarming phenomena. On the 7th, 8th and 9th of December, one was felt each day; and also on the 16th of January and 1st of April. (Mawson, 1943: 65)

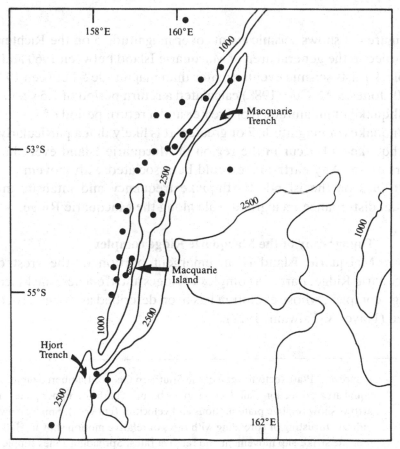

Figure 4.1. Locations of seismic events (●) greater than magnitude 5 in the vicinity of Macquarie Island, 1960–80. Drawn from unpublished data, Bureau of Mineral Resources, Canberra. Mercator Projection. 1000, 2500, . . . = contours in fathoms. 1 fathom = 1.83 m.

Table 4.1. *Earthquakes over magnitude 5 recorded in area shown in Figure 4.1 between 1924 and 1980.*

Year	Events	Year	Events	Year	Events	Year	Events	Year	Events
1924	1	1940	1	1966	6	1971	1	1977	2
1932	1	1943	1	1967	1	1972	3	1978	1
1935	1	1963	2	1968	1	1973	4	1979	2
1939	1	1965	1	1969	1	1974	2	1980	9

Source of data: Bureau of Mineral Resources, computerised earthquake data file.

Note: The apparent increase in seismic activity over the period shown may reflect advances in instrumentation rather than an actual increase in frequency of earthquakes.

Figure 4.1 shows seismic events over magnitude 5 on the Richter scale recorded in the general area of Macquarie Island between 1960 and 1980. Table 4.1 lists seismic events greater than magnitude 5 between 1924 and 1980. Jones & McCue (1988) calculated a return period of 1.5 years for an earthquake of magnitude 6 or greater and a return period of 7.5 years for earthquakes of magnitude 7 or greater. It is likely that a particularly large earthquake will occur in the region of Macquarie Island every hundred years or so. Any earthquake could be associated with movement along fault-lines on the island. Earthquake frequency and intensity indicate crustal disturbance on a grand scale along the Macquarie Ridge.

Topography of the Macquarie Ridge complex

Macquarie Island is an emergent portion of the crest of the Macquarie Ridge, part of a complex of ridges and trenches, the Macquarie Ridge complex, whose evolution has been described as extremely complicated (Hayes & Talwani, 1972).

Figure 4.2. Plate tectonic map of the Southern Ocean. Lambert azimuthal equal area projection. Subduction zones barbed (barbs on upper plate). Large arrows show relative plate motions and velocities (cm yr^{-1}). Small arrows indicate thrusting or spreading with rates of relative motion shown. Half arrows indicate strike-slip movement and relative rates. Spreading zones stippled. Marine magnetic anomalies 6 and 8 shown as dotted lines with numbers. Modified from Circum-Pacific Project, Plate Tectonic Map of the Circum-Pacific Region. Antarctica Sheet, scale 1:10000000, map centre point 20°S, 165°W, 1982.

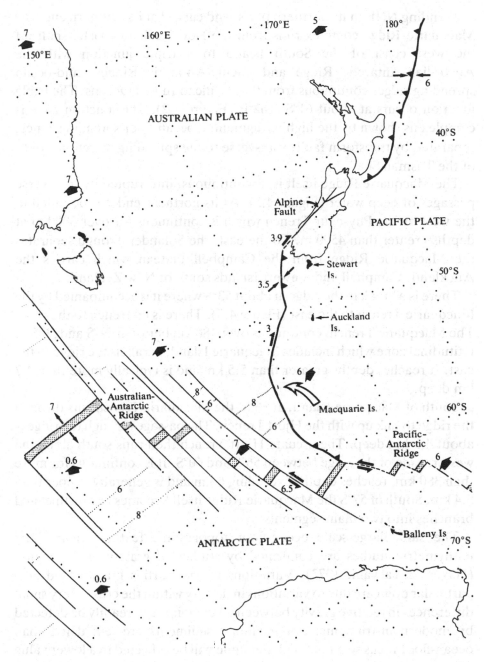

Trending NNE in its northern parts, and curved at its southern end, the Macquarie Ridge complex runs from the New Zealand coastal shelf off the west coast of the South Island to a triple junction with the Australian-Antarctic Ridge and Pacific-Antarctic Ridges, mid-ocean spreading ridges continuous from the Pacific to Indian Oceans. The triple junction occurs at about 61°S, 162°E (Figure 4.2). The junction zone is complex as shown by the highly fragmented ocean-floor spreading zones, separated by transform faults transverse to the spreading direction, south of the Tasman Sea.

The Macquarie Ridge itself is discontinuous, interrupted by transverse passages of deep water (Figure 4.3). At its northern end it is bounded to the west by the Puysegur Trench which is continuous for over 300 km at depths greater than 4500 m. To the east, the Solander Trough separates the Macquarie Ridge from the Campbell Plateau which carries the Auckland, Campbell and Stewart Islands south of New Zealand.

There is a break in the ridge at about 53°S where it is accompanied by the Macquarie Trench to the east (Figure 4.3). There is no trench to the west. The Macquarie Trench, continuous over 480 km between 51°S and 56°S, a latitudinal zone which includes Macquarie Island, parallels the ridge to the east. It reaches depths greater than 5.5 km and is typically more than 4.7 km deep.

South of Macquarie Island, at 56°S, the Macquarie Trench cuts through the ridge to link up with the Hjort Trench. The passage through the ridge is about 4.5 km deep. The arcuate Hjort Trench then runs south along the western side of the ridge between 56°S and 60°S. It is continuous for more than 380 km, reaches depths exceeding 6 km and is generally deeper than 5.4 km. South of 57°S the Macquarie Ridge itself becomes less distinct and branches into two main segments.

The same large-scale ocean-floor topographic features defined by bathymetric studies are confirmed by studies of gravity in the region (Hayes & Talwani, 1972). Variations in the Earth's gravity field at a particular point are due to variations in density within the crust. Very small differences in relative gravity between given points can readily be detected by modern instruments. For example, sediments are less dense than ocean-floor rocks so a thick sediment pile will be reflected in a lower value of gravity than a neighbouring point of the same elevation and latitude beneath which only igneous rocks are present. Variations in gravity, known as gravity anomalies, are greatest over ocean trenches. Gravity 'lows' are generally associated with oceanic trenches and gravity 'highs' with ridges parallel to them.

Tectonics

Studies of bathymetry, gravity anomalies and earthquake data all fit Macquarie Island's position at the boundary of two major tectonic plates.

The Macquarie Ridge complex is part of the boundary between the Australian and Pacific tectonic plates (Figure 4.2). The south-western margin of the Pacific plate is formed of oceanic ridges and trenches.

Tectonic activity along the Macquarie Ridge complex has probably been long-lived, judging from the fact that the Pacific plate has had a roughly constant direction of motion for the last 3–5 million years (Pollitz, 1986). Islands may have emerged and foundered repeatedly along the crest of the Macquarie Ridge throughout its history.

The Macquarie Ridge complex shows a mixture of extensional, compressional and transcurrent elements in the sea-floor. Seismic and other geological data are consistent with convergence of the Australian and Pacific plates about a pole of rotation to the south of New Zealand. The northern section of the ridge complex is an area of ENE-directed under-thrusting of the Tasman Sea floor (on the Australian plate) beneath the Pacific plate (Johnson & Molnar, 1972). Further south, the motion changes from thrust to strike-slip and then to normal faulting. Predominant motion along the plate boundaries seems to be lateral strike-slip at 3–5 cm yr^{-1} while other features indicate incipient subduction in some areas (Figure 4.2, Table 4.2).

Hayes & Talwani (1972) considered that a small amount of crustal extension may have occurred between 50°S and 57°S but cautioned against too ready an acceptance of the idea. Later studies (Williamson & Johnson, 1974) indicate crustal thickening, detected by gravity studies, towards the

Table 4.2. *Depths and magnitudes of earthquakes along Macquarie Ridge complex.*

Date	Location	Depth	Magnitude	Classification[a]
12 May 1963	57.5°S 159.4°E	44 km	6.2	TF or SF or NF
12 Sept. 1964	49.1°S 164.3°E	33 km	6.9	TF
8 Nov. 1964	49°S 163.7°E	33 km	5.6	TF
2 Aug. 1965	56.2°S 158.2°E	33 km	6.7	SF
25 May 1966	52.9°S 160°E	33 km	6.8	TF
20 Sept. 1967	49.8°S 163.4°E	30 km		TF

[a] NF = normal fault; TF = thrust fault; SF = strike slip fault.

Compiled from Johnson & Molnar (1972); Banghar & Sykes (1969); Sykes (1967).

Macquarie Ridge from the west between 50°S and 56°S, suggesting a compressional regime. This interpretation is supported by studies of first motions of earthquakes in the region (Jones & McCue, 1988). Such compression, together with crustal downwarping in the Macquarie Trench, may indicate incipient subduction of the Pacific beneath the Australian plate in the vicinity of Macquarie Island.

Williamson & Johnson (1974) interpret the Macquarie Trench, to the east of the Macquarie Ridge, as representing the boundary between the Australian and Pacific tectonic plates. This interpretation differs from that shown in Figure 4.2 where the boundary is shown to the west of Macquarie Island, following the line of the Hjort Trench.

Earthquake foci along the Macquarie Ridge complex are all shallow (less than 100 km deep), indicating that the complex is not a major subduction zone. Table 4.2 shows the location and depth of various earthquake foci along the ridge complex. Figure 4.1 shows seismicity in the Macquarie Island section of the complex.

Dating and general geology of Macquarie Island rocks

The rocks of Macquarie Island were formed by crustal accretion during sea-floor spreading at the Antarctic-Australian spreading ridge. Uplift of the block on which the island rides has been accompanied by both rotation and tilting of island segments with respect to each other (Williamson, 1978). At least some of the tilting and rotation of blocks is recent (Duncan & Varne, 1988).

On the basis of palaeomagnetic studies, Williamson (1978; 1988) decided that the island is made up of three main blocks. The northern and southern blocks have been tilted about an axis roughly perpendicular to the ocean-floor spreading lineations in the region. The central block is tilted to the north about an axis parallel to the sea-floor spreading lineations. The primary orientation, before tilting and rotation, was one of horizontal lava flows and vertical dykes, the dykes parallel to the sea-floor spreading pattern.

Palaeomagnetic studies of Macquarie Island rocks have given ages of 27–30 million years (Butler, Banerjee & Stout, 1975). Williamson (1988) described the island's rocks as having been formed at the spreading ridge during the Oligocene. Williamson regards the apparent palaeomagnetic pole from the island as consistent with the island's rocks having formed around the time of magnetic anomaly 7 (27 million years ago – Upper Oligocene) (Harland *et al.*, 1982).

Table 4.3. *Ages of Macquarie Island rocks.*

Age (millions of years)	Dating method, rock, site
0.01 (1)	14C, diatomite, Skua Lake
Middle to Late Pleistocene: *Emergence of island allows formation of subaerial deposits*	
Pliocene (2)	Palaeontological, interstitial oozes, North Head
9.7 (3)	K/Ar, pillow lava, Mawson Point
11.5 (3)	K/Ar, pillow lava, Pyramid Peak
Early to Middle Miocene (4)	Palaeontological, interstitial sediments, Pyramid Peak
27 (5)	Palaeomagnetic match with magnetic anomaly 7

Sources:
(1) Selkirk, Selkirk, Bergstrom & Adamson (1988).
(2) Varne, Gee & Quilty (1969).
(3) Duncan & Varne (1988).
(4) Quilty, Rubenach & Wilcoxon (1973).
(5) Williamson (1988).

Radiometric age determinations on pillow basalts yield dates significantly younger than those derived from palaeomagnetic studies and also show that not all the island's rocks were formed at the same time. Age determinations on relatively unaltered pillow basalts give crystallisation ages for basalts at Mawson Point and Pyramid Peak of 11.5 million years, and 9.7 million years for pillow basalts at North Head (Duncan & Varne, 1988) (Table 4.3).

Varne, Gee & Quilty (1969) suggested that interstitial *Globigerina* oozes in the pillow lavas were probably Pliocene in age and pointed out that the association of pillow lavas with interstitial *Globigerina* ooze, hyaloclastite and greywacke is characteristic of ocean-floor deposition at depths of 2 to 4 km. They regarded the island as raised to its present level since the Pliocene as a result of faulting and concluded that it is probably still rising, being the first to recognise beach deposits at 200 and 100 m altitude (Chapter 5). They also used observations from Macquarie Island to discuss processes occurring along the axes of mid-ocean ridges which

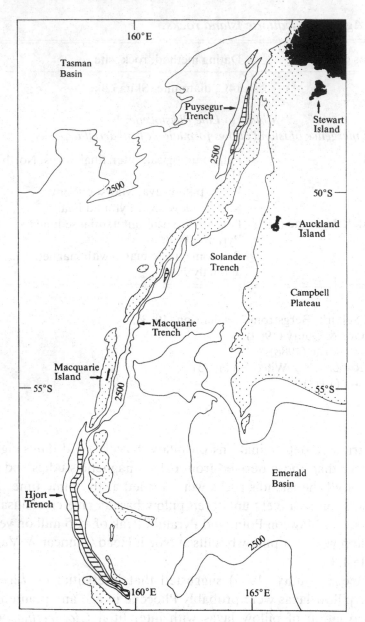

Figure 4.3. Generalised bathymetry of the Southern Ocean showing the Macquarie Ridge–Trench complex. Mercator projection. Land shown black. Areas 0–1500 fathoms stippled; 2500 fathom contour labelled. Areas deeper than 3000 fathoms hatched. 1 fathom = 1.83 m. Modified from Hayes & Talwani (1972).

lead to linear magnetic anomalies in ocean basins. Details of such processes were not fully known at the time.

Palaeontological studies of the interstitial oozes were made by Quilty, Rubenach & Wilcoxon (1973). They pointed out that the oozes and lavas are essentially coeval and dating of the oozes should thus provide a minimal age for the basalts. The interstitial oozes contain foraminifera, radiolaria, pteropods and coccoliths in an assemblage which suggests an Early (or perhaps Middle) Miocene age for the oozes and pillow lavas. Some reworked Cretaceous forms are also present.

On the basis of both structural relations and quality of rock preservation, Quilty, Rubenach & Wilcoxon (1973) suggested a gradation in age in pillow basalts from south of Mawson Point to North Head, basalts from North Head being the younger. This prediction has been borne out by radiometric dating.

Datings proposed for rocks from Macquarie Island are shown in Table 4.3. A mid-to-Late Miocene Age for the pillow lavas seems most likely.

Rock types on the island

The first general study of the island's geology was made by L. R. Blake during his time there with the Australasian Antarctic Expedition. His extensive field notes were later worked up and published by Mawson (1943). The island's geology has recently been surveyed by Christodoulou, Griffin & Foden (1984), Crohn (1986), and Duncan & Varne (1988). The rock sequences mentioned here are described fully in Christodoulou, Griffin & Foden (1984).

The volcanic and sedimentary rock sequence

The volcanic and sedimentary sequence exposed on Macquarie Island is basically made up of three components: pillow basalts and basalt flows, a series of sediments which owe their origin to the volcanic sequence (volcaniclastic sediments), and sediments of marine origin deposited in the interstices of the volcanic and volcaniclastic rocks.

Figure 4.4 shows that the overwhelming bulk of the island is made up of pillow basalts (Figure 4.5) with associated sediments and interspersed flows of massive basalt. The volcanic and associated sedimentary sequence makes up most of the southern two-thirds of the main island mass and also occurs in the isolated bulk of Wireless Hill. Basalt pillows differ considerably in size and shape. They have glassy margins and contraction cracks indicating rapid cooling of lava as it was extruded on the ocean floor. Flows

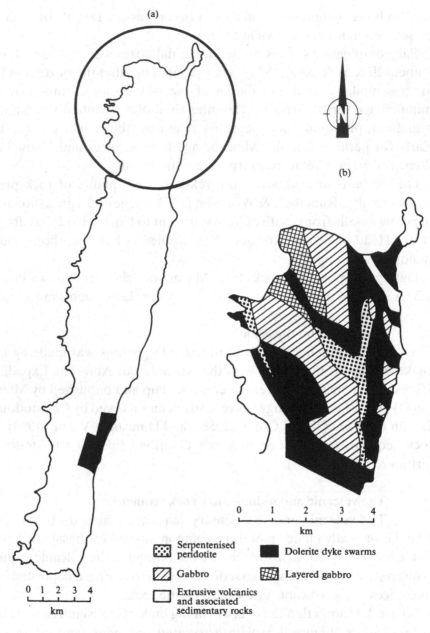

Serpentenised peridotite

Dolerite dyke swarms

Gabbro

Layered gabbro

Extrusive volcanics and associated sedimentary rocks

Figure 4.4. Generalised geological map of Macquarie Island, modified from Christodoulou, Griffin & Foden (1984) and Duncan & Varne (1988). (a) Whole island; (b) Details of northern part of island.

Figure 4.5. Knoll of pillow lava immediately south of Pyramid Peak. Human figure gives scale.

of massive basaltic lava up to 6 m thick occur among the pillows. No feeder dykes to pillows or massive basalts are exposed. The pillows may have built up mostly from material supplied to them through pillows whose exteriors chilled and ruptured to release liquid magma to form other pillows.

Within the lava pile there is a gradation from zones of closely packed pillows without interstitial material to zones of loosely packed pillows with up to 30% interstitial material. In order of decreasing abundance, the following rock types occur:

(a) pillow basalts,
(b) paraconglomerates,
(c) lithicwackes and reddish mudstones,
(d) massive lava flows, and
(e) hyaloclastic breccias and oozes.

Paraconglomerates or conglomeratic mudstones are basically mudstones whose fine-grained matrix contains larger fragments of rock. Paraconglomerates exposed on Macquarie Island presumably formed as a result of submarine turbidity flows and submarine slides. The lithicwackes are soft rocks of fine texture derived from disintegrated basaltic rocks. The term reddish mudstone is self-explanatory. Hyaloclastic breccias consist of fragments of volcanic glass which are not welded together.

Oozes are present on the floors of all oceans, containing the remains of microscopic marine animals. The oozes exposed on Macquarie Island are highly calcareous; up to 49% $CaCO_3$ (Mawson, 1943). They are fine-grained and generally highly indurated. In some localities on the island it is possible to recover microfossils for study.

Dyke swarm complexes

These occur as two small exposures north of Lusitania Bay on the east coast, as well as forming a major element in the complex of various rock types exposed in the northern third of the main island.

The main characteristic of the sheeted dyke complexes is that dolerite dykes intrude among other dolerite dykes so that extensive sheets of parallel dykes are formed. The chilled margins of the dykes face the same direction, characteristic of being formed as a part of sea-floor spreading (Gass, 1982). The width of individual dykes is very variable. The central zones of the dykes are porphyritic, containing phenocrysts (large crystals) embedded in a finer-grained crystalline matrix. The most common phenocrysts are plagioclase feldspars composed of silicates of aluminium, calcium, potassium and sodium. The central porphyritic zone of the dykes grades outwards to a zone lacking phenocrysts and then a micro-crystalline chilled margin. Size of phenocrysts decreases from the centre of the dyke towards the margin. The structure of the dykes and distribution of phenocryst phases within them suggest slow cooling of the magma which formed them and a relatively plastic environment of formation. All the dykes of the dyke swarm complex have been affected by alteration since their original intrusion.

These dykes differ petrographically from those which intrude the pillow basalts and basalt flows (Varne & Rubenach, 1972). The latter are only rarely porphyritic and are more finely grained with basaltic textures.

Layered sequence of troctolites and gabbros

Gabbros are the coarse-grained equivalents of basalts and dolerites, formed at greater depths within the crust. Troctolites are a type of gabbro which contain no pyroxene minerals and whose plagioclase fraction is between 50% and 70% anorthite, a plagioclase mineral with high calcium content. Pyroxenes are anhydrous silicates of magnesium, iron, calcium, aluminium, sodium and lithium in various combinations and ratios.

At Eagle Point, the main rock types of the layered sequence, which is extensively faulted, are plagioclase-containing dunites, troctolites made up of alternating olivine-rich and plagioclase-rich layers, and small exposures of anorthosite, a coarse-grained rock which is more than 90% plagioclase. Layering within the troctolites is not of the type typically due to settling within a magma.

Along the shores of Half Moon Bay, troctolite exposures are discontinuous and intruded by dolerite dykes which cut across the layering pattern at various angles. Olivine gabbros are exposed near Handspike Point. In the gabbros there has been progressive substitution of olivine by pyroxene and in some exposures the olivine has been entirely replaced.

Peridotites

The peridotites exposed on Macquarie Island, mainly at Langdon Bay and Eagle Point, are harzburgites, which contain a mixture of olivine and a pyroxene mineral, enstatite.

Small exposures of harzburgite on the plateau, originally mistakenly interpreted by Blake as glacial erratics (Mawson, 1943), are serpentinised. Serpentine is the main alteration product of olivine and pyroxene minerals. Serpentinisation is a metamorphic process caused by hydrothermal action on ultrabasic rocks rich in olivine and pyroxene.

The outcrops of harzburgite show small-scale faulting along curved surfaces. Chrysotile, a form of asbestos, is developed along fault planes. There are major fault zones in the harzburgite at Unity Point. At Langdon Bay, dykes and veins of gabbro are present in the harzburgite.

Gabbroic rocks

Several bodies of gabbro are exposed on the plateau and along the east coast. The outcrop around Island Lake is the largest. Small dolerite dykes are common within the gabbro. On the western slopes of the plateau is a zone of cataclasite, formed along fault zones as a result of fracture, rotation and recrystallisation of mineral crystals. Slickensides, polished and scratched surfaces on a fault zone produced by friction between opposing sides of a fault, are present in the cataclasite. The cataclastic zone appears to reflect upward movement of harzburgite relative to the gabbroic rocks.

Gabbroic exposures along the east coast are nested in dolerite dykes but contacts between the two rock types are poorly exposed. Gabbroic bodies on the plateau are highly altered and weathered and occur among dolerite dykes.

Macquarie Island as an ophiolite complex

Recognition of the fact that Macquarie Island represents an ophiolite complex, a slice of oceanic crust, in relatively pristine condition has generated considerable geological interest (Varne, Gee & Quilty, 1969; Varne & Rubenach, 1972; Williamson & Rubenach, 1972; Butler, Banerjee & Stout, 1975; Williamson, 1978; Griffin & Varne, 1980; Cocker, Griffin & Muehlenbachs, 1982; Christodoulou, Griffin & Foden, 1984; Crohn, 1986). The best-known ophiolites so far subjected to detailed geological study are fragments of ocean-floor emplaced on continental margins, suffering inevitable massive deformation in the process (e.g. the Troodos massif of Cyprus and the Samail nappe of Oman). The processes whereby ocean-floor material ends up as part of continents rather than disappearing at subduction zones are not yet fully understood.

For a general discussion of ophiolites and their theoretical significance see Gass (1982) and Kerr (1983).

Macquarie Island must be one of the few places on Earth where it is possible to walk about relatively easily on a slice of ocean-floor material surrounded by thousands of kilometres of open ocean. The island is made up of a series of faulted blocks, representing different crustal layers, tilted with respect to one another and cross-faulted on all scales, providing several partial sections through oceanic crust.

A schematic section through the original column of rock before it was disrupted and raised above sea-level is shown in Figure 4.6. The geological map of the island (Figure 4.4) can be interpreted in terms of this schematic section.

There is a stratigraphic sequence of:
1. an upper volcanic and sedimentary sequence,
2. a sheeted dyke complex represented by the dyke swarms,
3. a sequence of layered troctolites and gabbros,
4. harzburgites,
5. gabbroic bodies whose exact position within the sequence is hard to determine.

The depths within the crust at which various portions of the sequence formed are indicated on Figure 4.6.

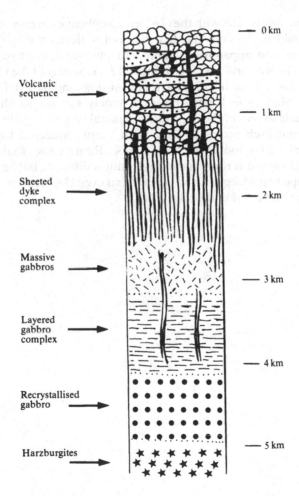

Figure 4.6. Schematic section through original, hypothetical rock column of Macquarie Island ophiolite complex, prior to uplift and faulting. Modified from Cocker, Griffin & Muehlenbachs (1982).

Cocker, Griffin & Muehlenbachs (1982) concluded that the island's rocks have undergone metamorphism during their formation on and under the sea-floor by interaction with circulating seawater to depths of at least 4–5 km. They postulate two seawater circulation regimes. In the upper regime, basalts were weathered by cold (<20 °C) seawater circulating through the rock pile in a circulation system with high water:rock ratios. At deeper levels, seawater at high (300–600 °C) temperatures circulated at low water:rock ratios. Crohn (1986) disputes this interpretation of seawater circulation regimes.

The geological importance of Macquarie Island is neatly summed up by Kerr (1983: 1309):

> One way to deal with the obvious complexities of most ophiolites would be to find one that has not suffered the bashing and alteration apparently required to put ocean crust onto continents. R. Varne, B. J. Griffin, and J. A. Jenner of the University of Tasmania think they have a promising candidate. It forms part of Macquarie Island, the last, lonely outpost south of New Zealand. All of the appropriate crustal layers are there, and the island itself seems to have been simply squeezed toward the surface like toothpaste from a tube. Remoteness, foul weather, and rugged terrain make field studies difficult, but researchers hope that Macquarie will provide an even clearer view of at least one section of mid-ocean crust.

5

Geomorphology and Quaternary history

Just as the geological origin of Macquarie Island differs from that of other subantarctic islands, so do its landscape and the processes shaping it. Neither subaerial volcanoes nor glaciers, both features of other subantarctic islands, are to be found on Macquarie Island. Its present landscape results from the interplay of faulting, uplift, sea-level changes, erosion and periglacial processes.

Faulting

Macquarie Island is a block of ocean-floor material on the eastern edge of the Australian plate, raised from great depth in mid-ocean and subject to frequent, quite severe earthquakes (Chapter 4). Fault movements on a monumental scale must be responsible for its uplift and for that of the whole of the Macquarie Ridge.

The position of the major faults which outline the blocks of ocean-floor material on which Macquarie Island sits remain unknown. The linear east coast of the island gives every indication of being fault-determined. Other major faults probably lie offshore. The embayed nature of the west coast is not so neatly controlled by faulting but again major faults probably lie offshore. The island's vegetation and extensive cover of loose material make it difficult to map geologically as faults and contacts between rock units are often obscured.

Three major faults or groups of faults crossing the island have been mapped by the juxtaposition of rocks of different lithologies (Figure 4.4; Christodoulou, Griffin & Foden, 1984; Duncan & Varne, 1988). The northernmost of these faults crosses the Isthmus near the fuel depot of the present ANARE station, and separates Wireless Hill from the main part of the island. The second extends from Langdon Bay on the west coast to Sandy Bay on the east, running in part along the southern side of the Finch Creek valley. The third crosses the island from Sandell Bay on the west

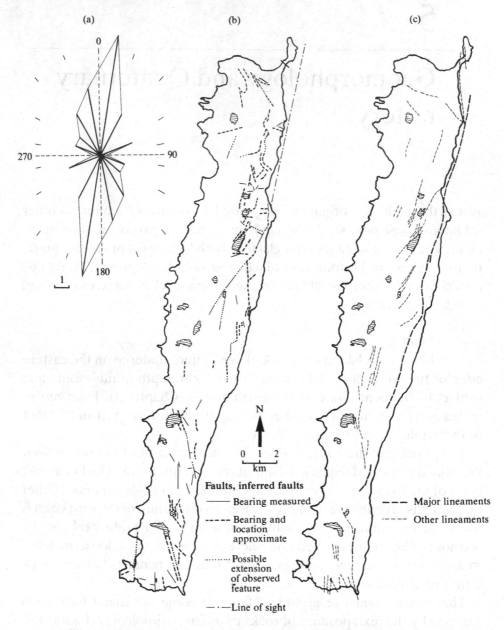

Figure 5.1. Macquarie Island, showing orientations of faults and inferred faults. (a) Rose diagram showing frequency of fault orientations in 10° sectors. (b) Map based on field observations and aerial photograph interpretation by D. Adamson, 1987. (c) Map based on Figure 2, Ledingham & Peterson (1984). Redrawn with permission from Royal Society of Tasmania.

coast north of Major Lake, to the east coast between Waterfall and Lusitania Bays.

The landscape of the island is intensely influenced by faulting. Faults close to the east coast determine coastline and clifflines. The orientations of numerous valleys and creeklines, and the locations of most lakes, are fault controlled. Small-scale faults and joints in bedrock are exploited by erosion to form gullies.

For much of its length, the eastern coast of the island parallels a lineament known as the Brothers Fault-line (Ledingham & Peterson, 1984), which is a complex system of faults (Figure 5.1). An upthrust wedge associated with this fault system forms a series of ridges marking the eastern edge of the island plateau between Saddle Point and Brothers Point (Figure 5.2). In the southern part of the island, a series of faults with the same general alignment is clear between Mt Jeffryes and Hurd Point.

Along the eastern side of the island there are small fault lines oriented roughly E–W and NW–SE (90° and 135°) but it is clear that NNE trending lineaments dominate the eastern seaboard (Figure 5.1(a)). The eastern coastline itself is presumably defined by a series of faults of similar orientation to the Brothers system of faults. This structural control results in a fairly straight coastline with very few bays.

Figure 5.2. A ridge associated with the Brothers Fault-line forms the eastern edge of the plateau near Brothers Lake.

'Brothers Fault-line has been used as a convenient name for one of the most impressive and clearly defined lineaments on the island' (Ledingham & Peterson, 1984: 231). It has been mapped as a series of lineaments running just inland or just offshore along most of the island's east coast. It

Figure 5.3. A rampart-like ridge with eastern side upthrust 2–3 m runs along the eastern coastal cliff about 50 m below the plateau rim between Brothers Point and Brothers Lake.

should be regarded as a zone of parallel and subparallel active normal faults, not as a single major fault.

A variety of fault-associated features occurs along its length. Between Nuggets Point and Sandy Bay, Brothers Fault-line probably defines the outer edge of the coastal rock platform. To the north, along the same alignment, Judge and Clerk Islands can be seen (on the occasional clear day) from a vantage point near Brothers Point. This alignment, when extended southwards, passes through the Bishop and Clerk Islands, making a total distance of about 80 km over which the alignment can be followed. Between Brothers Point and Green Gorge there is a coastal ridge whose western (that is, inland) slope is covered with scree formed from fault-breccias. Between Brothers Point and Brothers Lake, a rampart-like ridge (Figure 5.3), with the eastern side upthrust 2–3 m, runs along the coastal cliffs about 50 m below the plateau rim. At Green Gorge, an uplifted block, or series of small blocks, flanks the bay (Figure 5.4).

In the northern wall of the gully of a tributary of Sawyer Creek, between Green Gorge and Pyramid Peak, a zone of fault breccia can be seen in the western side of the elevated coastal block. The fault strikes N–S and dips approximately 55° west. The eastern block appears to have been upthrust.

Figure 5.4. A series of uplifted blocks flanks the northern and western shores of the bay at Green Gorge.

This observation is contrary to Ledingham & Peterson's (1984) interpretation of an ESE dip for the Brothers Fault-line. It is also clear at two other points that the dip of the component faults is westerly. The notch on the skyline and the slope on the westerly face of the rampart-like fault scarp shown in Figure 5.3 have a westerly dip: an easterly dip at this site would be parallel to the slope of the cliff face and would produce no visible topographic feature.

At the south-eastern corner of the island there are several sets of faults, each set with different orientation. The orientations of these fault-lines have been mapped from aerial photographs (Figure 5.1(b)). One of the fault systems matches in orientation the Brothers Fault-line. A fault from another orientation-set forms a particularly spectacular lineament above Hurd Point at the southern end of the plateau. A scarp, about 20 m high at its southern end, diminishing towards the north, bears 145°–150° (approximately NNW–SSE) (Figure 5.5). The slope on the scarp of 30° with the top more easterly and the direction of the gully running from scarp to sea-level down the cliff are consistent with a westerly dip.

Overall, the prominent faulting along the eastern side of the island (including the Brothers Fault-line) forms an alignment slightly east of north although in detail there is a zig-zag of individual faults oriented N to NNW and NE to NNE.

In contrast to the east coast, on the west coast there is no obvious lineament comparable to the Brothers Fault-line visible on the island. Here, structural control takes a different form. The major western faulted margin of the island block presumably lies offshore, west of the marine terrace. The indented bays and the raised marine terrace have been formed by vigorous erosion of the western coast of the island. Many headlands parallel lineaments visible on the plateau surface. In eroding the western shore, the sea would have exploited any joints or faults in the bedrock of the cliffs. Bays would have been eroded in small units bounded by joints and faults, and hence have margins matching them in orientation.

Many lakes on the island appear to be fault controlled (Ledingham & Peterson, 1984), either lying in lineament-bounded depressions or formed by fault-damming of drainage lines. Lake Ainsworth, on the plateau close to the southern coast of the island, lies in a fault-bounded depression (Figure 3.2). The fault-line which forms its western margin bears 180°, and extends as a dry gully from the plateau margin down to sea-level.

Lake Ifould once drained south-eastwards towards Green Gorge. Upthrusting along its south-eastern margin diverted its outflow towards

Figure 5.5. Fault scarp at southern plateau margin above Hurd Point.
Australian Antarctic Division photograph, S. Brown.

Figure 5.6. Square Lake drains via a notch (arrow) in a fault scarp on its western side into Stony Creek.

the north-east (Figure 6.13). The course of its outflow creek appears to have been diverted on more than one occasion by block fault activity. It is now a tributary to Red River, flowing along the eastern side of a 100 m wide upthrust block which trends approximately 30°. This block was incorrectly described as a kame terrace (Colhoun & Goede, 1974), but Ledingham & Peterson (1984) identified it as a fault.

Between Sandy Bay and Mt Blair, near the eastern margin of the plateau, four small lakes occupy a curving valley bounded to the east by a ridge which follows a curving fault-line. An active fault scarp forms the easterly edge of each lake. The lakes, described by Löffler, Sullivan & Gillison (1983) as glacial, are fault-controlled.

Square Lake (Figure 5.6) is formed by a fault on its western margin which dams a wide mire-filled valley. Stony Creek drains Square Lake via a notch in the fault scarp.

Uplift

Since the island is composed of ocean-floor material (Chapter 4), dramatic uplift has clearly occurred. Several lines of evidence point to uplift of the island surface above sea-level as relatively recent.

During the process of uplift, every level on the island must, at some stage, have been at sea-level. As a result of the fluctuations in global sea-level during the Quaternary, some parts of the island would have been at or near sea-level more than once. Uplift could have occurred either as a continuous process or in a series of lurches of activity separated by periods of little or no upward movement. During net uplift of the island, there may have been some periods of overall downward movement, as well as the movement of some blocks downward with respect to others.

Around the seashore today, waves break on pebble and cobble beaches and against rocky cliffs. In the past, beaches would have formed when a suitable part of the coastline was at sea-level for a sufficient length of time. Preservation of relict beach deposits (horizontal or gently seaward-sloping terraces strewn with well-rounded pebbles or cobbles at levels now clearly above sea-level), first noted by Varne, Gee & Quilty (1969), provide evidence for a particular part of the island having been at sea-level. In estimating time of formation of a particular beach level, both global sea-level changes with time and net uplift rates of the island have to be considered.

Figure 5.7 shows horizontal slices of the island at 100 m intervals which can be likened to maps of the island at four arbitrarily selected times during the emergence of the island. First, isolated small islets would have appeared, followed by an archipelago, a few large islands, and finally, the present island. Since the island is composed of a number of blocks which are more likely to have moved individually than as an island-sized unit, Figure 5.7 is certainly an oversimplification of events. The outlines shown in Figure 5.7 are inaccurate as they do not take account of the major changes in shape and area of the island resulting from vigorous marine erosion as the island rose and world sea-level fluctuated.

Beaches could have formed on any part of the present island when it was a coastline. In theory, the oldest beaches would now be the highest in altitude if uplift of all blocks making up the present island had proceeded in unison. In practice, it is likely that different blocks moved at different rates, and that beaches of similar ages are now at different elevations.

Ledingham & Peterson (1984) mapped as raised beach deposits level terraces with pebble or cobble scatter or lag now at altitudes ranging from 10 to 270 m. Of particular interest in unravelling the geomorphic history of the island is dating the time at which particular beaches were at sea-level and calculating the rate of uplift to their present altitudes.

Several estimates have been made of rates of uplift operating on the island over the past few thousand years. Assuming immediate peat

(a)   (b)   (c)   (d)

297 000	268 000*	165 000**	80 000
123 000	100 000	77 000	45 000
80 000	62 000	48 000	32 000

Figure 5.7. Notional uplift sequence for Macquarie Island. Fine outline shows
present Macquarie Island. Areas above sea-level of the time shown black.
Times (in years before present) at which each map would have represented
Macquarie Island are shown for each of three assumed uplift rates: 1.5, 3.25
and 5 mm per year. * and ** represent calculated times of first emergence
above the ocean. With changing sea level, parts of the island would have been
resubmerged before ultimately rising above sea level approximately * 210 000
and ** 125 000 years ago.

formation on any area lifted above wave influence, and basing their calculations on radiocarbon dates from the base of peat deposits overlying beach cobbles at Green Gorge, Colhoun & Goede (1973) calculated a maximum rate of beach terrace uplift of 4.5 mm per year. McEvey & Vestjens (1973) had previously obtained radiocarbon dates for fossil penguin bones at Sandy Bay. From these dates, Colhoun & Goede calculated a minimum rate of uplift of 1.5 mm per year. They concluded that the raised beach terraces at Bauer Bay and Green Gorge were raised above sea-level between about 6000 and 2000 years ago. It is impossible to know whether uplift was continuous or occurred in a series of sudden movements and whether movement was entirely vertical or involved tilting of the island. Extrapolating from calculated uplift rates and the altitude of the island they postulated that the island first emerged in the middle to late Pleistocene.

Bergstrom (1985) made a detailed study of peat deposits in the Green Gorge basin. Based on palynological evidence and the presence of hair from elephant seals and fur seals at the base of the deposits, she considered that the site studied was close to sea-level 8500 years ago and had been uplifted at a rate of 3 to 4 mm per year. A later, more accurate measurement of the present altitude of the site allows recalculation of the rate of uplift for the site as 2 mm per year. From studies of a mire site in the Green Gorge basin, she calculated an uplift rate of 6 mm per year. Recalculation following altitude remeasurement gives an uplift rate for the site of 5 mm per year, remarkably good agreement with Colhoun & Goede's (1973) calculated rate of 4.5 mm per year for a site a few hundred metres away.

Selkirk, Selkirk & Griffin (1983) studied a deposit of peat overlying a substantial marine beach deposit now at an altitude of 100 m on the western side of Wireless Hill. Detailed palaeontological study indicated that the basal layers of the deposit must have been formed close to (within about 20 m of) sea-level where the vegetation was affected by the activities of elephant seals. Basal layers of the deposit gave a radiocarbon age of 4880 ± 90 years BP (SUA-1527) or, with correction (Klein *et al.*, 1982)., 5580 ± 260 years BP, implying that uplift of the site to its present altitude of 100 m must have been rapid (14.5 mm per year). Such an uplift rate is much higher than that calculated for the beach terraces and the Green Gorge area on the main part of the island. However, Wireless Hill is separated from the rest of the island by a major fault (Figure 4.4) and is probably a small fault-bounded block, acting essentially independently of the rest of the island, and whose uplift proceeds at a different rate.

Figure 5.8. Benches in the coastal ridge at Lusitania Bay are likely to be edges of fault blocks.

In addition to mapping raised beaches bearing deposits of well-rounded pebbles and cobbles, Ledingham & Peterson mapped 'terrace levels: areas of uniform height which may be of beach origin' (1984: Figure 1 legend). Some of these 'terrace levels' may be of beach origin, but many are not.

In the vicinity of Lusitania Bay, sub-horizontal benches top the cliffline between 100 and 200 m. Although any remnant beach material on their surfaces would be buried under peat slipped from slopes above, making beach deposits difficult to locate, it seems clear from the irregular shape and uneven surfaces of the benches that they are most likely to be edges of fault blocks (Figure 5.8).

To the north and south of the beach at Green Gorge, stepped benches are visible in the coastal ridge (Figures 5.4, 3.4). These benches are also associated with faulting and appear to be fault block edges.

A number of sub-horizontal benches occur at about 200 m asl at the top of the cliffs at the northern end of the island, on the southern shore of Hasselborough Bay. The plateau above these benches is strewn with well-rounded cobbles. The lowermost bench, at about 230 m asl, has an uneven sloping surface. The outer rim of this bench is bedrock, confining towards its western end several pools, one of which drains as a creek through a gap in the bedrock rim. The only rounded cobbles found on this bench are at the base of a spill from the cobble-mantled bench above. Two smaller benches above are strewn with cobbles of similar dimensions and state of weathering as those on the plateau. The orientation of the outer margins of the benches is approximately 125°, matching one of two predominant orientations of lineaments in the region (Figure 5.1). We interpret this area at the northern margin of the plateau as the north-western extension of Ledingham & Peterson's (1984) Beach 3, disrupted after or during uplift to its present elevation by differential movement along fault block margins.

Sea-level changes and coastal erosion

Uplift rates calculated for parts of Macquarie Island have all been based on consideration of features younger than about 8000 years, during most of which time global sea-level has remained essentially unchanged. In interpreting features older than this, and in considering the time of emergence of the island above the sea, sea-level changes over time must be taken into account.

Late Quaternary global sea-level changes are known in general terms (Chappell, 1983; Chappell & Shackleton, 1986). If we assume that the rates of island uplift calculated for the Holocene were average rates throughout the mid to late Quaternary, we can infer relative altitudes of island surface and sea level at various past times. Uplift rates of between 1.5 mm yr^{-1} and 6 mm yr^{-1} (revised to 5 mm yr^{-1}) have been published for sites in the northern half of the main island, and 14.5 mm yr^{-1} for Wireless Hill.

Mt Hamilton, the highest point on the island, is 433 m asl and, but for erosion, would presumably have been a little higher, say 450 m asl. For the minimum 1.5 mm yr^{-1}, the mean 3.25 mm yr^{-1} and maximum 5 mm yr^{-1} assumed uplift rates, we calculate that Mt Hamilton would first have appeared above present sea-level 300 000, 140 000 or 90 000 years ago when sea-level was respectively approximately 30, 25 and 50 m below its present level (Figure 5.9). Mt Hamilton would, then, have appeared above sea-level at the time of 315 000, 160 000 or 94 000 years ago. Since the rates are assumed rates, the few additional years are of little significance in calculating the approximate date of the island's first emergence. Previous estimates of the time of the island's emergence from the sea are

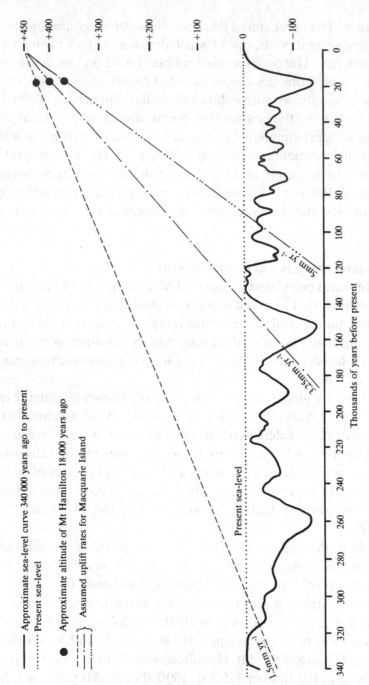

Figure 5.9. Sea-level curve from 320000 years ago to the present (redrawn from Chappell, 1983) on which assumed uplift rates for Macquarie Island have been superimposed to allow estimation of altitude of Mt Hamilton at past times.

Metres above present sea-level

——— Approximate sea-level curve 340 000 years ago to present
·········· Present sea-level
● Approximate altitude of Mt Hamilton 18 000 years ago
– – – Assumed uplift rates for Macquarie Island

consistent with these calculations: Colhoun & Goede (1973) considered the 100 000 year age associated with the 4.5 mm yr^{-1} uplift rate was too recent to have allowed development of the erosional topography. They favoured the middle Pleistocene age associated with a 1.5 mm yr^{-1} uplift rate. Ledingham & Peterson (1984) suggested a late Quaternary emergence.

The uplift rates of 1.5, 3.25 and 5 mm per year and the Quaternary sea-level curve (Figure 5.9) have been used to place tentative ages on the 'time series' maps of Macquarie Island during its emergence from the sea. The notional sequence shown in Figure 5.7 is drawn from the present contour map of the island. Below each map are shown times when, according to each of the three uplift rates, this map would have represented the above-sea parts of Macquarie Island. The small islets shown in Figure 5.7(a) are those parts of the island now above the 400 m contour. We calculate that this much of the present island would have been above the sea 297 000, 123 000 or 80 000 years ago if the island had been rising at a rate of 1.5, 3.25 or 5 mm yr^{-1} respectively. Similarly, Figures 5.7(b) (300 m contour), 5.7(c) (200 m contour), and 5.7(d) (100 m contour) show times when they would have represented the island.

As the island emerged above sea-level its shores were subject to marine erosion (Figure 5.10(a)). Beaches with well-rounded pebbles or cobbles were formed on some shores. Preservation of some of these by island uplift has been discussed above.

Vigorous wave action against sea cliffs led to undercutting at sea-level, falling away of the cliff edge, and hence progressive retreat of the cliff margin. Cliff retreat progressively truncated features on higher parts of the island above the cliffline. Within the last 300 000 years there have been considerable changes in sea-level (Figure 5.9). The severity of marine erosion and the preservation of coastal features was affected by the balance between uplift of the island surface and changes in global sea-level.

Various scenarios can be proposed for the relationship between sea-level and uplift of the island. If the emerging land were close to sea-level for a considerable period, marine planation would occur (Figure 5.10(b)). If sea-level were rising faster than the island, progressively higher levels would be subject to marine erosion (Figure 5.10(c)). Then a bench with a surface sloping seaward would be formed on which debris from erosion of the cliffs would at first accumulate. Depending on its nature and the length of time it was subject to marine influence, rubble would be pounded into

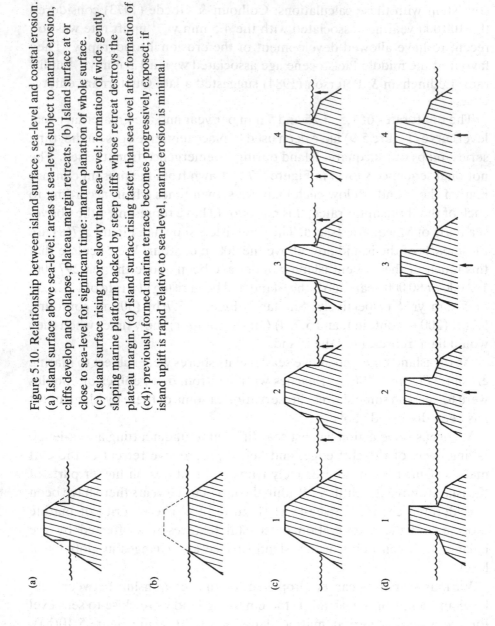

Figure 5.10. Relationship between island surface, sea-level and coastal erosion. (a) Island surface above sea-level: areas at sea-level subject to marine erosion, cliffs develop and collapse, plateau margin retreats. (b) Island surface at or close to sea-level for significant time: marine planation of whole surface. (c) Island surface rising more slowly than sea-level: formation of wide, gently sloping marine platform backed by steep cliffs whose retreat destroys the plateau margin. (d) Island surface rising faster than sea-level after formation of (c4): previously formed marine terrace becomes progressively exposed; if island uplift is rapid relative to sea-level, marine erosion is minimal.

smaller pieces including rounded cobbles, pebbles and sand which would accumulate *in situ* or be washed elsewhere. The slope of the bench would depend on the relative rates of uplift and sea-level rise. Finally, if the island were rising faster than sea-level, coastal features resulting from marine erosion would be progressively raised above sea-level and preserved from further erosion. Little coastal erosion would result if uplift markedly outstripped sea-level change.

Many landforms on the island can be interpreted in these terms, and, by comparison with Figure 5.10, estimates of their age made. Some examples follow.

Wireless Hill

From palaeontological study of peat overlying a raised beach on the western side of Wireless Hill (Beach 1 of Ledingham & Peterson, 1984), Selkirk, Selkirk & Griffin (1983) calculated an average uplift rate of about 14.5 mm yr^{-1}. For the period between 16 000 and 8000 years ago sea-level was rising world wide at approximately 15 mm yr^{-1} (about 120 m in 8000 years). If Wireless Hill had been rising at a constant rate, its surface would have been close to sea-level for the 8000 years between 16 000 and 8000 years ago before rising above the waves to its present altitude. Marine planation during this long period at about sea-level is probably the cause of its flat top (Figures 5.10 and 2.7).

West coast marine terrace and cliff retreat

The origin of the west coast marine terrace between sea-level and 10 to 15 m asl can be explained in terms of marine erosion and uplift (Adamson, Selkirk & Colhoun, 1988). Between 16 000 and 8000 years ago, sea-level rise of about 15 mm yr^{-1} was much more rapid than uplift of the island (at between 1.5 and 5 mm yr^{-1}). Erosion at the coast occurred as in Figure 5.10(c), resulting in formation of a sloping underwater terrace partly covered with debris from the eroded cliffs. Much of this was no doubt lost into deeper water or moved along the shore. The terrace sloped up towards the land and terminated in steep sea-cut cliffs. After about 6000 years ago when sea-level stabilised, no further terrace would have formed, but continued uplift of the island gradually lifted the terrace surface and its landward cliffs above sea-level (Figure 5.10d).

Vegetation now mantles the emerged part of the terrace, but well-rounded cobbles can be located on it in ponds and drainage lines where peat is absent. The emerged terrace is between 200 and 1500 m wide to the north of Bauer Bay, and slopes 1.5° to 4° down towards its seaward edge. Using the model outlined above, Adamson, Selkirk & Colhoun (1988) predicted that the terrace is between 1 and 3 km wide, i.e. between two and five times its so-far-emerged width. They estimated the seaward edge of the terrace to be 70 to 100 m below sea-level. The few soundings along the north-west coast of the island (Division of National Mapping, Australia, 1982) are consistent with these estimations. Detailed soundings will test the model.

Associated with formation of the marine terrace would have been destruction of a strip of island plateau equal in width to the marine terrace. Creeks which may have had gentle gradients to sea-level were truncated by progressively retreating cliffs, and now fall sharply from the plateau edge to the marine terrace, for example Flat Creek, the creek from Flynn Lake and the creek from the small lakes south of Major Lake.

Some lakes formerly close to the edge of the plateau were drained when the plateau margin was cut back by marine erosion. Skua Lake is a remnant of a formerly much larger Palaeolake Skua which extended westward beyond the present cliff margin. The evidence is clear. On top of the 180 m cliff overlooking the ocean just west of Skua Lake ancient lakebed deposits are exposed extending about 9 m above and 5 m below the present level of Skua Lake. These deposits can be followed eastwards in the ridges flanking the south-western end of Skua Lake. A layer, 4.6 m above the base of the deposit, containing plentiful freshwater diatoms and leaves and stems of *Myriophyllum* sp. and aquatic mosses, gave a radiocarbon age of 12 470 ± 280 BP (Beta-20165). A layer rich in *Myriophyllum* and aquatic mosses, 2.2 m from the top of the deposit (i.e. about 7 m above the present level of Skua Lake), gave a radiocarbon age of 8620 ± 170 BP (Beta-20166) (Selkirk, Selkirk, Bergstrom & Adamson, 1988). Palaeolake Skua evidently occupied a wide shallow basin to the south-west of Mt Ifould from at least 14 000 years ago (taking calibration of the radiocarbon dates into account) to about 9000 years ago when it was drained by collapse of the cliffs to the west into what is now Sellick Bay. Remnants of the formerly extensive lake remain as present-day Skua Lake and a series of ponds.

On the east coast of the island just north of Nuggets Point, a lakebed deposit topped by modern peat forms a 4.5 m vertical exposure at about 30 m asl in the coastal cliff. A layer close to the base of the deposit, containing leaves and stems of *Myriophyllum* and aquatic mosses, pollen and freshwater diatoms, gave a radiocarbon age of 9400 ± 400 BP (SUA-1894). A layer at the top of the lacustrine material, just below the overlying modern peat, gave a radiocarbon age of 6850 ± 100 BP (SUA-2158). This deposit is also interpreted as a remnant of a lake, Palaeolake Nuggets, which formerly extended eastwards beyond the present cliffline. The lake was in existence on the site by 10 000 years ago and was drained by cliff retreat due to marine erosion some time soon after 7500 years ago.

Western plateau

On the western margin of the plateau, above cliffs sloping steeply down to a marine terrace below, are two relatively level areas at about 200 m asl. The more northerly, flanked to the east by Mt Ifould, includes the shallow Skua Lake and the site of the former Palaeolake Skua. The more southerly, flanked to the east by Mt Gwynn and a northern spur of Mt Martin, contains Major Lake (Figure 3.5). On the western margins of each of these areas, relict beach deposits have been described (Beaches 8 and 9 of Ledingham & Peterson, 1984). These flattish areas of plateau may represent a marine terrace cut during the period between 150 000 and 135 000 years ago when sea-level was also rising rapidly for a long period (Figures 5.9, 5.10(c)). The western sides of Mts Ifould, Gwynn and Martin would then represent the now eroded western sea-cliffs of the island of that time, debris from whose erosion formed the beach cobbles now visible along the top of the present sea-cliffs. It is not feasible to estimate the westerly extent of the marine terrace of that earlier time as the surface has been modified by subaerial erosion, possible tilting and differential uplift, and the phase of severe marine erosion between about 16 000 and 8000 years ago which destroyed a significant strip of the plateau as it formed the present marine platform. Faulting has certainly modified the flattish plateau surface west of Mt Gwynn, including the basin containing Major Lake. The best exposure of well-rounded cobbles near Major Lake is beside the seaward end of a fault across the north-east corner of Major Lake.

Glaciation

The importance of faulting in determining landform as well as major geological boundaries on the island has only recently been recognised (Ledingham & Peterson, 1984). Blake, surveyor and geologist with the Australasian Antarctic Expedition on Macquarie Island 1911–14, recognised the separation of rock types in the northern plateau from those

of the rest of the island by a large fault (Mawson, 1943) and postulated a
submarine fault or series of faults to account for the steep sea-floor
gradient off the east coast. Blake considered glaciation to have been the
major force moulding the island's landforms. Faulting evidently seemed
more significant to Mawson than to Blake. Mawson commented, in
connection with the north-west corner of the island, that faulting is likely
to have 'played a part in arriving at the present topography' (Mawson,
1943: 66).

The respective roles of glaciation and faulting in moulding landforms on
Macquarie Island have been the subject of considerable debate. The
earliest view, developed by Blake, and followed by Mawson, was that the
island had been massively glaciated during the last glacial maximum,
totally over-ridden by a thick ice-sheet which accumulated to the west of
the present island. This ice-sheet was believed to have covered the plateau
and the rest of the island down to sea-level, leading to the development of
features such as the flat top of Wireless Hill. Blake recognised features
such as glacial striae, ice-polished rocks around the margins of lakes,
moraine-dammed lakes, over-deepened rock basins occupied by lakes,
roches moutonnées and a mantle of till covering much of the present island
surface.

Taylor (1955a) followed Blake's ideas and concluded that the island's
present plant and animal inhabitants must be relative newcomers – a biota
which arrived by long-distance dispersal from other land masses to
colonise an island left bare of vegetation by a receding ice-sheet.

The idea of a major over-riding ice-sheet has since been rejected.
Interpretations of the extent and severity of glaciation vary and the reality
of any glaciation at all has been questioned.

Colhoun & Goede (1974) believed there had been limited glaciation
with local plateau, valley and cirque glaciers accumulating in depressions,
valleys and basins on the plateau, glaciers occupying about 40% of the
present island surface. Glaciation of this type would have left extensive
refugia for island organisms on the present island mass as well as on areas
exposed by a then-lower sea-level.

Löffler & Sullivan (1980) postulated a glaciation of much greater extent,
at least on the northern third of the island, basing much of their interpretation on aerial photographs. They made no observations on the southern
part of the island. They considered that most of the northern part of the
island had been glaciated, leaving only small ice-free refugia in the west
and north. They described east-facing cirques along much of the eastern
plateau margin and a line of moraines between them and the coast. They

concluded that most of the plateau must have been capped by an almost continuous thin layer of ice and that solifluction processes operating since the last glacial maximum had considerably modified old glacial terrain. Areas which they believed had escaped glaciation were the steep coastal escarpments, the coastal platform on the west coast and the shelf which would have been exposed by lower sea-levels.

Ledingham & Peterson (1984) questioned the earlier, glacially-based interpretations of the island's geomorphic history, suggesting that rapid tectonic uplift associated with substantial block faulting and development of many major structural lineaments has been the most important element in development of the present landscape (Figure 5.1). They described raised beaches and terrace lines, possibly of marine origin, at many locations on the island in areas formerly regarded as glaciated and suggested that the shape and distribution of many of the plateau lakes and various lineaments observable on the island could be explained by processes other than glaciation. They considered, however, that certain features were still best explained by very minor glaciation. Adamson, Selkirk & Colhoun (1988) question even these in the Bauer Bay–Sandy Bay region.

This view is not universally accepted. Crohn (1986) believes that 'glaciation occurred, affecting most or all of the present plateau surface', although he also considers that tectonic and erosional forces have been important.

In the light of recent studies, Colhoun now holds different views from those he published in 1973.

> Demonstration of the widespread importance of faulting, and interpretation of some gravels on the northern part of the island as marine rather than glacigenic in origin, strongly support an interpretation of relatively recent uplift for the island. Previous records of glaciation now seem quite uncertain, and there remains a need to establish whether or not the island was at any time and in any part glaciated.
>
> (E. A. Colhoun, pers. comm.)

In considering the likelihood of Pleistocene glaciation on Macquarie Island it would be helpful to know the elevation of the island above sea-level at the time, as well as the likely snowlines. At the three uplift rates assumed above, any level on the island would, 18 000 years ago, have been about 25, 60 or 90 m lower than it is now except that, as sea-level of the time was about 120 m lower than at present, any part of the island would have had an elevation above sea-level of the time similar to or slightly greater than at present. For example, Mt Hamilton, now 433 m asl was, 18000 years ago, 525, 495 or 460 m asl (Figure 5.7).

Today there is no permanent snow on Macquarie Island, one of few subantarctic islands without it. South Georgia rises to 2934 m, Heard Island to 2745 m, Iles Kerguelen to 1978 m, Marion Island to 1230 m, whereas Macquarie Island is only 433 m high. In addition to its low altitude, Macquarie Island's extreme maritime climate with almost continuous drizzle and strong winds would promote snow-melt and minimise accumulation. Although snow may fall at any time of year, it does not lie for more than about a week. Colhoun & Peterson (1986) estimate the modern snowline on Macquarie Island at approximately 500 m asl.

At the time of the last worldwide glacial maximum about 18000 years ago, Macquarie Island's altitude above sea-level was probably little different from today's, as described above. World temperatures are estimated to have been about 4–6 degrees colder than at present, making Macquarie Island temperatures similar to those on Heard and Macdonald Islands today. Macdonald Island, whose highest point is 212 m asl, is also free of permanent snow. Mass balance measurements for the Vahsel Glacier on Heard Island indicate an equilibrium line about 200 m asl (Allison & Keage, 1986), agreeing with an earlier estimate of snowline elevation (Lambeth, 1951). The highest parts of Macquarie Island now have elevations within a few hundred metres of that permitting accumulation of permanent snow and ice. It remains a matter for conjecture whether, during the last glacial period, conditions on highest hills of the island allowed perennial snow to accumulate. If so, the lack of geomorphic evidence for glaciation suggests accumulation was very limited. Whether the whole of the island was still below snowline or not, it seems to us that glaciation has not played an important part in sculpting the island's landscape.

Erosion

Erosion is very active on Macquarie Island, yet there have been few studies of erosion processes there. Soil creep, a continuous process with material gradually moving downslope under gravity, was thought by Griffin (1980) to be the major erosion mechanism on the island, although landslip, water and wind erosion are also counted as important. Rapid slope failure is also clearly important, but quantitative studies are few.

The regular precipitation and low evaporation which are characteristic of subantarctic climates result in a permanently saturated mantle/bedrock interface. This facilitates downslope movement of material, both slowly as soil creep and, when a suitable trigger occurs, rapidly as landslips. Landslips are relatively common, particularly on the steep coastal slopes. Sometimes they are triggered by periods of heavy rain, when presumably the weight of sodden peat becomes too great to remain on very steep slopes, or the sodden peat can no longer support heavy tussocks at precarious angles on steep slopes. Earthquakes, common on this seismically active island, are another trigger to landslips (Griffin, 1980).

Although landslips leave areas bared of vegetation and sometimes of surface peat, the areas revegetate quite quickly in the moist environment. An area photographed in 1958, at that time bared of vegetation by landslip, was completely covered with short grassland vegetation when rephotographed in 1980 (Selkirk, Costin, Seppelt & Scott, 1983).

Differences in the early stages of revegetation are related both to the quantity and nature of substrate left after the initial disturbance, and the extent of rabbit grazing on the site. Rabbits affect revegetation by grazing new growth on dominant plant species and reexposing shallow peat by digging (Scott, 1985) (Figure 6.10). It has been suggested that selective grazing by rabbits of stabilising species in grassland, and weakening of an area by burrowing activities, increased the incidence of landslips and accelerated the erosion of steep slopes (Taylor, 1955a; Costin & Moore, 1960). Although selective grazing by rabbits has considerably changed the floristic composition of grassland and herbfield vegetation, rabbit activity is not a major erosive force on the island (Griffin, 1980; Selkirk, Costin, Seppelt & Scott, 1983).

In addition to triggering rockfalls and landslips, seismic activity triggers the formation of peat cracks. We have observed peat cracks several centimetres wide and several metres long in two situations; along faultlines and at the heads of valleys. After a substantial earthquake in February 1980, visible cracks formed parallel to the major fault-lines along the coastal ridge north of Red River, and on the western side of Wireless Hill plateau. Cracks up to 10 cm wide developed in *Poa foliosa* tussock-covered slopes on the eastern side of Razor Back Ridge.

Curving cracks in surface peat are frequently seen around the heads of wide shallow valleys such as that to the south of Brothers Lake. These peat cracks, too, are likely to have been initiated in saturated peat by seismic activity or by slow downslope movement of the saturated peats. Subsequent slumping contributes to the headwards erosion of the valley, and the accumulation of material on its wide, shallow floor.

Water erosion is responsible for deepening and widening creek gullies both in peaty sites on the plateau (Selkirk, Costin, Seppelt & Scott, 1983) and in rocky sites such as Gadgets Gully on the coastal slopes. In feldmark areas on the plateau, surface water flowing across unvegetated areas washes fines downslope, leaving a gravel lag.

Between Bauer and Sandy Bays, a thin sand sheet (at least 6000 to 7000 years old) mantles the plateau surface (Adamson, Selkirk & Colhoun, 1988). Flowing surface water has washed sand away to expose an area of rounded beach cobbles (Beach 5 of Ledingham & Peterson, 1984), to expose bedrock and to form gullies.

Wind is an active erosive agent in the Bauer Bay area. Contemporary wind polish on rock can be found close to sea-level about 100 m inland of high tide level on the sandy shores of Bauer Bay. Ancient wind polish and wind-eroded grooves occur on bedrock exposed at the plateau margin high above Bauer Bay. It is clear that wind-driven sand is not active at this site today since mosses and lichens colonise the ventifacted rock surfaces. The long axes of the wind-eroded grooves are oriented at 310°, parallel to the present prevailing wind direction, indicating that the dominant wind direction has not altered since their formation (Adamson, Selkirk & Colhoun, 1988).

Inland of Bauer Bay, wind action is today eroding areas on the plateau sand sheet to form blow-outs, elongate trough-shaped hollows (Adamson, Selkirk & Colhoun, 1988). The long axes of the blow-outs are parallel to the prevailing strong wind direction. The sandy floors of the blow-outs continue to be eroded by wind and water, while the outer margins and down-wind tip where blown sand is deposited are progressively stabilised by creeping plants such as *Acaena magellanica* and *Ranunculus biternatus*.

Loose sand, blown and washed eastwards across the plateau, has accumulated in *Acaena*-covered hummocks in the upper Finch Creek valley (Adamson, Selkirk & Colhoun, 1988). Although aeolian landforms are unusual on subantarctic islands, they are well known from cold wet areas of Europe.

Periglacial phenomena

Although climatic records for Macquarie Island indicate that average daily minimum, mean and maximum temperatures are all above 0 °C, an average of 51 days per year have air temperatures below zero, and 99 days per year have surface ground temperatures below zero (Löffler, 1983).

As meteorological data are recorded at 6 m asl, and temperatures are approximately one degree lower for each 100 m increase in altitude, occurrence of ground frost at plateau sites is likely to be significantly more

Figure 5.11. Stone stripes resulting from frost action run downhill on the west side of North Mountain. Notebook (15 × 9 cm) gives scale.

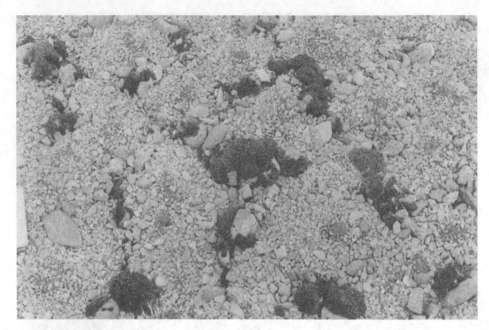

Figure 5.12. Stone polygons on level ground on the north side of North Mountain.

frequent. Löffler (1983) estimated there would be approximately 170 days with frost at 200 m asl. Precipitation averages about 900 mm per annum at the Isthmus meteorological station, and 42% more than this on the plateau at its northern end (Mallis, 1988), so moisture for needle ice formation on the plateau is abundant. Ground temperatures at depths greater than 10 cm approximate air temperatures, so soil frost formation is restricted to the surface 10 cm or so (Löffler, Sullivan & Gillison, 1983).

On bare ground on the plateau, stone sorting resulting from frost action is common: stone stripes, small stone polygons and sorted downslope movement of gravel. All these periglacial features are of small dimensions, indicating that they result from short-term (daily) rather than long-term (seasonal) frost cycles (Troll, 1960).

Stone stripes run downslope on surfaces sloping 2° to 20° (Figure 5.11). Parallel lines of gravel are separated by finer material forming mounds 2–3 cm high and 10–25 cm but most commonly about 20 cm wide (Löffler, Sullivan & Gillison, 1983; Peterson, Scott & Derbyshire, 1983). The degree of sorting varies from site to site.

Stripes sometimes run around cushions of vegetation on a slope, sometimes terminate in stripes of vegetation running across the slope, called windward terraces by Taylor (1955b). Fines, transported downslope by surface wash, will accumulate in the upslope edge of the vegetation in windward terraces, providing additional substrate and nutrients for the plants' growth and contributing to the upslope migration of these terraces. Feldmark terrace form and vegetation is discussed in Chapter 6.

Small sorted stone polygons, about 20 cm in diameter, occur on more-or-less level ground in a few places. The larger stones which form the rim of each polygon often bear clumps of moss such as *Andreaea* spp. (Figure 5.12).

Vegetation history

Pollen from subantarctic plants appears not to be widely dispersed (Barrow, 1978). Larger plant fragments such as seeds and leaves are commonly quite locally deposited. Palaeontological examination of terrestrial peat and lacustrine deposits on Macquarie Island have been useful in reconstructing past vegetation and environmental history of a particular site.

Although Bunt (1956) described pollen from lignites which he interpreted as belonging to a preglacial Macquarie Island flora, his deposits were undated, and more recent studies have failed to confirm his interpretation. Studies of terrestrial peat dated at up to 8500 radiocarbon years old, and of lacustrine deposits up to 13 200 radiocarbon years old, have revealed no evidence for an island flora different from today's (Selkirk, Selkirk & Griffin, 1983; Salas, 1983; Bergstrom, 1985; Selkirk, Selkirk, Bergstrom & Adamson, 1988). These studies have also provided evidence for the long-distance transport, in very low numbers of pollen grains from Australia and New Zealand to the island.

6

The island's vegetation

Geographical isolation and prevailing climate markedly affect vegetation of subantarctic islands. They support neither trees nor shrubs. The tallest plants are grass tussocks of *Poa* or *Parodiochloa* up to 1.5 m tall. Bryophytes, lichens and low-growing vascular plants such as cushion plants, sedges, grasses and forbs are important. In some ways vegetation of subantarctic islands is similar to northern hemisphere tundra (Bliss, 1979). Arctic tundra is characterised by semiprostrate and dwarf woody shrubs with an understorey ground cover dominated by mosses and lichens. Macquarie Island vegetation lacks woody plants.

All subantarctic islands have diverse seabird and marine mammal faunas, some land-birds, higher insects, spiders, earthworms and molluscs. Close to sea-level, seabirds and marine mammals can have marked influences on vegetation. Grazing, burrowing, trampling and manuring by native or introduced animals all affect plant growth.

History of botanical studies on Macquarie Island

The first collections of plants from Macquarie Island were made by sealers or sailors early in the nineteenth century and forwarded to the then Superintendent of the Sydney Botanic Gardens. The collection which included eight vascular species and some stray bryophytes was forwarded to W. J. Hooker at Kew in 1824.

After a visit to the island in 1880 (Chapter 2), J. H. Scott described its general appearance as

> barren in the extreme. There is not a tree or shrub, and what vegetation there is has a great degree of sameness, long stretches of yellowish tussock, with occasional patches of the bright green *Stilbocarpa polaris*, or of the peculiar sage green *Pleurophyllum*. These, with the rich brown mosses near the hill tops, are all that strike the eye on looking at the island

> from the sea . . . the interior . . . shows the rocky tops of the hills blown
> perfectly bare by the winds, and fissured by the frosts . . .
>
> (Scott, 1883: 486-7)

He noted the affinity of the island's flora with that of New Zealand and
listed eight mosses and one hepatic, the first record of bryophytes from the
island. Scott's vascular plant collections were later described by
Cheeseman (1919).

In 1894, A. Hamilton spent 13 days on the island collecting vascular
plants and published a list of 32 species (Hamilton, 1895). He considered
three species to be naturalised aliens – *Stellaria media, Cerastium triviale*
(now known as *C. fontanum*) and *Poa annua*. Cheeseman (1919) commented that, from Hamilton's visit, the most important result

> was the totally unexpected discovery of three new species of grasses –
> *Deschampsia penicillata, Poa hamiltoni* and *Festuca contracta*. Up to that
> time no one had even suspected that Macquarie Island had an endemic
> flora of its own, and the establishment of that fact not only marked an
> important advance, but also compelled a rearrangement of all previous
> views on the history and development of the vegetation of the island.
>
> (Cheeseman, 1919: 14)

The belief that the island supported endemic species stimulated interest
in its flora. Vascular species then believed to be endemic have since been
found to be more widespread, consistent with the idea that, since the
island is geologically young, it would have few or no endemic species. This
idea may require rethinking in the light of Orchard's (1989) segregation of
the endemic *Azorella macquariensis* from the widespread *Azorella selago*.

A. Hamilton's son, H. Hamilton, was a member of the Australasian
Antarctic Expedition 1911–14 (AAE) on Macquarie Island (Chapter 2).
His collections added four new species to the island's known flora. Dodge
(1948) identified lichens collected by H. Hamilton, and those collected
during a very brief visit to the island in 1930 by the British, Australian and
New Zealand Antarctic Research Expedition (BANZARE).

Most of the botanical collections made by BANZARE were destroyed
during the second world war. Sir Douglas Mawson, leader of both the
AAE and BANZARE, noted:

> The botanical collections of our late enterprise (the B.A.N.Z.A.R.
> Expedition of 1929–31) made on subantarctic islands of the Indian Ocean
> were extensive. All but the lichens were forwarded for report to the
> British Museum where they suffered destruction during one of London's
> bombing raids of the recent war . . . lichens, together with those of the
> earlier expedition . . . were despatched to America . . . the main body of

the lichens collected on Macquarie Island during the occupation of 1911–1914 was included by mistake in the general botanical collection forwarded and subsequently lost, as a result of war-time bombing.

(Dodge, 1948: preface)

Since establishment of the ANARE station on the island, extensive collections of vascular plants have been made. Taylor (1955a) published a detailed account of the island's vascular flora, vegetation and soils. Intensive studies of grasslands and herbfields have been made (Jenkin, 1972, 1975; Jenkin & Ashton, 1979) and an annotated atlas of the vascular flora has recently been published (Copson, 1984).

Lichen collections were studied by Dodge (1948, 1968, 1970) and Dodge & Rudolph (1955). Bunt made a number of mycological collections (Bunt, 1965). Clifford (1953) published an account of the moss floras of both Heard and Macquarie Islands, listing 40 species from Macquarie. Recent extensive field surveys have shown that Macquarie Island supports 45 vascular plant, more than 80 moss, 50 hepatic and 100 lichen species (Seppelt, 1977, 1981; Filson, 1981; Seppelt, Copson & Brown, 1984) (Appendixes 1, 2, 3).

Little work has yet been done on the terrestrial and freshwater algal flora of the island. Bunt (1954b) made a preliminary survey of terrestrial and freshwater diatoms in the course of a comprehensive survey of the microbiology of island soils. Evans (1970) and Croome (1984) expanded the list of known freshwater algae (Appendix 5). The marine algal flora has only recently received attention. As a result of a survey made during 1977–78, the first extensive list of algae fringing the island has been published (Ricker, 1987) (Appendix 6).

Vegetation formations

Taylor (1955a) recognised five vegetation formations: grassland, herbfield, fen, bog and feldmark. Within each formation he recognised one or more alliances, associations and sub-associations:

Grassland (wet tussock) – on all steep coastal slopes up to ăltitudes of about 330 m, in immature river valleys and inland slopes protected from severe wind, on flats of the coastal terrace except where the water table is very high and occasionally on upland flats protected from severe wind.

Herbfield – on slopes and flats subject to moderate winds and in all areas where wind velocity is not too high; in sheltered valleys, on slopes up to a maximum altitude of 380 m and on raised coastal terraces.

Fen – on valley floors, on the plateau and in small patches on raised coastal terrace; occurs where the water table is at or slightly above the ground surface and where the water is neutral or alkaline due to contact with underlying basic rocks and mineral soil.

Bog – occurs where the water table is at or slightly above the ground surface and where the water is acid and low in soluble salts due to contact with underlying peat soil.

Feldmark – occurs in all areas subject to high wind velocities; covers most of the island above 180 m, occasionally occurring as low as 90 m.

Our studies on the island show that Taylor's vegetation classification is sound but oversimplified. An expanded classification for the whole of the subantarctic region (Smith, 1984) is useful but does not adequately describe the vegetation of individual subantarctic islands. We have modified Taylor's vegetation classification as follows:
  (i) Grassland is divided into tall tussock grassland, dominated by *Poa foliosa* tussocks, and short grassland, which includes those areas which are meadow-like and dominated by species of *Agrostis, Luzula, Uncinia* and *Deschampsia* or *Festuca*.
  (ii) We recognise a fernbrake community, dominated chiefly by *Polystichum vestitum*.
  (iii) Mire incorporates Taylor's fen and bog where the water table is at or near the surface.
In the section which follows we have adopted a zonal approach to description of the island's vegetation, discussing the vegetation of the coastal zone, beach slope and raised coastal terraces, coastal slopes and uplands, rather than the vegetation associations occurring in those areas.

Coastal communities
The coasts of Macquarie Island and plant and animal communities along its maritime fringe are strongly influenced by waves and salt-spray. Storms are common and the high mean wind velocity ensures steady deposition of salt-spray over the coastal areas.

Composition and structure of near-shore communities is determined by many factors: topography, which determines degree of exposure to waves and wind; proximity to the sea; whether or not a beach is present and the nature of any beach; substrate stability and composition; surrounding vegetation and animal disturbance.

Figure 6.1. In the lichen zone above the high tide limit, species of *Verrucaria*, *Xanthoria*, *Mastodia*, *Lecanora* and the moss *Muelleriella crassifolia* are common. Large fronds of *Durvillaea* dominate the intertidal zone. Green Gorge.

Coastal rock communities

Marine algae occupy the intertidal zones (Figure 6.1). Above this is a lichen-dominated zone where species of *Verrucaria*, *Xanthoria*, *Mastodia* and *Lecanora* are common. The moss *Muelleriella crassifolia* often occurs in this zone, forming dense tufts on rocks and producing abundant capsules each year. Plants in this zone are often inundated by salt water in rough weather.

Dense cushions of *Colobanthus muscoides*, tufts of the grass *Puccinellia macquariensis*, two moss species and several lichen species occur above the lichen zone on coastal rocks (Figure 6.2). *Puccinellia* often grows in areas drenched by waves during exceptional storms. *Colobanthus* occurs

Figure 6.2. Maritime community, Hasselborough Bay. Tufts of the grass *Puccinellia macquariensis* grow amongst cushions of *Colobanthus muscoides*. Halophytic lichens and moss cover exposed rock.

above the area of wave influence, occasionally associated with *Crassula moschata* (Taylor, 1955a).

The zonation sequence in coastal rock communities is very variable. In many places *Puccinellia* is found seawards of *Colobanthus* but the sequence may be reversed or the two species intermixed.

Many coastal rocks are capped by short tussocks of *Poa foliosa*, often associated with plants of species of *Cotula, Stilbocarpa, Agrostis, Festuca, Cardamine, Acaena* and bryophytes such as *Hypnum cupressiforme, Metzgeria* spp., *Lophocolea bidentata, Macromitrium longirostre, Pottia heimii* and *Bryum dichotomum*.

A different zonation sequence occurs on steep beaches composed of large, wave-worn stones. Above the algal zone, lichens may extend to the upper level of the beach and abut dense stands of *Poa foliosa* which often have a seaward border of *Cotula plumosa, Poa annua, Colobanthus muscoides* or *Crassula moschata*. Above the algal zone, beach stones are often covered with species of the lichens *Verrucaria, Xanthoria, Caloplaca, Mastodia* or *Lecanora*.

Headlands, particularly those with large remnant rock stacks seaward of them, are vegetationally complex. Stacks are often capped with *Poa foliosa* tussocks which provide nest sites for burrow-nesting petrels, cormorants, rats and mice, and occasionally rabbits. In such areas there is often a well-defined vegetational sequence from a seaward lichen-dominated zone which may be mixed with vascular plants, species of *Colobanthus, Puccinellia, Crassula, Cotula* and the moss *Muelleriella crassifolia*. Other lichens (species of *Rinodina, Parmelia, Buellia, Cladonia, Pseudocyphellaria, Graphis, Ramalina*) and bryophytes such as *Macromitrium longirostre, Bryum dichotomum, Pottia heimii, Lembophyllum divulsum, Lophocolea bidentata* and *Metzgeria* spp., are found together with vascular plants such as species of *Isolepis, Agrostis, Stilbocarpa, Acaena, Cardamine* and *Callitriche*.

Soil or peat development over the irregular substrate, shelter from direct sea-spray provided by intervening rock stacks, and drainage play an important role in determining the complexity of the plant associations. Animals such as fur seals and nesting birds may locally influence the dominance of a particular species. For example, it is possible that the abundance of *Cotula plumosa* on coastal sea stacks at Handspike Point has been brought about by cormorant breeding and the presence of New Zealand fur seals which haul out in the area in late summer.

Beaches

Beaches on Macquarie Island are formed of sand, shingle or rounded cobbles. Beach profiles change regularly during storms and rough weather. Mobile beaches lack vegetation. Detached subtidal algae, particularly *Durvillaea antarctica* and *Macrocystis pyrifera*, are regularly heaped on the beaches. They decay rapidly due to the activities of bacteria and kelp fly larvae. Elephant seals, especially during the annual moult, often lie in the rotting kelp (Figure 9.3).

Cotula plumosa, Poa annua and, particularly where freshwater seepage or runnels occur, *Callitriche antarctica* and filamentous algae occur on the upper levels of sandy beaches. The mosses *Bryum argenteum, B. dichotomum* and *Pottia heimii* colonise more stable sand. On stony beaches,

Figure 6.3. Zonation of coastal vegetation: *Callitriche antarctica, Cotula plumosa* and *Poa annua* occupy the seaward fringe of dense *Poa foliosa* tussock grassland. A mixed *Poa foliosa – Stilbocarpa* community on the slope beyond is replaced by short grassland at its upper limits.

Cotula plumosa, Poa annua and *Callitriche antarctica* (and occasionally *Crassula moschata* and *Colobanthus muscoides*) form a fringe to maritime tussock grassland. The uppermost levels of most beaches support a tall tussock grassland of *Poa foliosa* (Figure 6.3).

Vegetation dominated by large tussock-forming grasses is characteristic of coastal vegetation of a number of cool temperate and subantarctic islands. On South Georgia, *Parodiochloa flabellata* is the tussock-forming species (Headland, 1984); on Macquarie Island, *Poa foliosa* forms tall tussocks, *Poa cookii* and *Poa litorosa* smaller tussocks. On Heard and McDonald Islands and Iles Kerguelen, the tussock-forming species is *Poa*

Figure 6.4. Wind-sculpted remnant tussocks on the beach at Bauer Bay.

cookii. Wace (1960: 480, Table 8) summarises the principal components of maritime tussock communities on various subantarctic and cool temperate islands.

On Macquarie Island, tall tussock grassland occurs as closed or open vegetation dominated by *Poa foliosa.* Mature tussocks may be 2 m or more high. Although the pedestals may be well separated the canopy is often closed and few vascular plants other than *Cardamine corymbosa, Stellaria decipiens* and *Epilobium* spp. are commonly found beneath the canopy. Tussock growth is comparatively rapid but decay of dead leaves is relatively slow: thus tussock stools are clothed with a dense covering of dead and decaying leaves and the ground below them is carpeted with litter.

The pattern of regeneration of *Poa foliosa* suggests a cycle of development through pioneer, building, mature and degenerate phases, with a gap phase involving small herbs, grasses, bryophytes or *Stilbocarpa polaris.* Regeneration of *Poa foliosa* is almost exclusively by rhizomes (Ashton, 1965). Growth rate is high despite the cool climate (Table 6.1). Seedling establishment, despite a high seed yield, is infrequent. Movement of animals, particularly penguins and elephant seals, between tussock pedestals may considerably hasten erosion of peat, accentuating the height of the pedestals.

Table 6.1. *Biomass and annual production in Macquarie Island grassland and herbfield.*

	Grassland[a]	Herbfield[b]
Altitude (m asl)	45	235
Above-ground biomass (g m^{-2})	3610	2693
Total biomass (g m^{-2})[c]	8148	2693
Living above-ground biomass (g m^{-2})	917	541
Total living biomass (g m^{-2})[c]	2579	1211
Estimated annual above-ground production (g m^{-2})[d]	3282	489
Estimated annual total production (g m^{-2})[d]	6954	1036

[a] Most abundant species at site = *Poa foliosa* and *Stilbocarpa polaris*.
[b] Most abundant species at site = *Pleurophyllum hookeri*.
[c] Total of above-ground and below-ground parts.
[d] Takes account of loss by decomposition, principally by fungal activity: calculated rate of dry weight disappearance (mg g^{-1} day^{-1}) for *Poa foliosa* = 2.74, *Pleurophyllum hookeri* = 1.99, *Silbocarpa polaris* = 6.52.

Data from Jenkin (1972). All figures mean of results from periodic harvests over 13 months.

Near colonies of royal, king and rockhopper penguins, *Poa cookii* tussocks are intermixed with those of *P. foliosa*. Although *P. cookii* shows lushest growth near penguin colonies where it must be influenced by increased nutrient levels, it is also found at a few sites distant from animal influence. *Poa annua*, *Callitriche antarctica* and *Cotula plumosa* are found where seals and/or penguins have an impact on the vegetation, colonising and stabilising peat surfaces damaged by animals.

Two areas of sandy beach, one at the southern end of the Isthmus, the other at Bauer Bay, have shown considerable change in vegetation cover over the years. Hamilton (1926: 7) referred to the sandy beach on the Isthmus as the

> only sand area on the island. It owes its existence to deflection of the prevailing westerlies by a rocky ridge, the sand being drawn up onto the lower slopes of the hills. The only vegetation on these sand slopes is an occasional patch of *Poa foliosa*. The plant formation is gradually being killed by the advancing sand.

The lower slopes on the eastern side of the ridge are now covered by dense *Poa foliosa*. Below, on the upper part of the beach, the sand has been colonised by a luxuriant carpet of *P. annua* which is now being invaded by *P. foliosa*. Little remains of the sandy slopes of Hamilton's time.

A further extensive area of sandy beach occurs at Bauer Bay on the west coast. Isolated *Poa foliosa* tussocks are scattered over the beach, their bases strongly sculpted by wind erosion (Figure 6.4). At Bauer Bay there is also a dune system, covered by *P. foliosa*. As with tussocks on the beach, tussocks in the dunes are heavily eroded by wind and wind-blown sand. Removal of sand is accelerated by elephant seals during their moulting period. Large volumes of sand are carried inland by wind towards elevated coastal terraces at the eastern end of the bay. Sand in the area is clearly quite mobile, and changing vegetation boundaries on it are common.

Raised coastal terraces

There are numerous areas of raised coastal terrace on the island, quite distinct from sandy or stony beaches (Chapter 5). They support a variety of vegetation formations whose composition is governed largely by influx of run-off water from the coastal slopes behind, rainfall and drainage capacity of underlying peats. The most complex vegetation associations occur on the extensive raised terrace on the north-west coast between Aurora Point and Hasselborough Bay.

Well-drained areas on upper beach slopes and raised coastal terraces support mixed stands of *Poa foliosa* and *Stilbocarpa polaris* (Figure 6.5). Taylor (1955a) described the *P. foliosa* association as incapable of growing in soil with a stagnant water table. *Poa foliosa* does, however, grow along streams. Such stream-side communities are particularly obvious where streams cross a coastal terrace (Figure 6.6). Stands of *P. foliosa* are found on small islands in some lakes including the saturated islands in Floating Island Lake (Figure 7.6).

Areas of *Poa foliosa* and *Stilbocarpa polaris* on coastal terraces provide transitory refuge for elephant seal pups in the post-weaning period and for adult seals during moulting. They also provide an important habitat for mice, rats, rabbits and wekas (Chapter 12), and are common breeding sites for solitary-nesting northern giant petrels, colonially-nesting southern giant petrels and for gentoo penguins.

Areas of short grassland, dominated by *Festuca contracta* often with subdominant *Luzula crinita* and *Agrostis magellanica*, occur on some of the more extensive coastal terraces, such as that near Half Moon Bay. Short grassland of *Poa annua* and *Agrostis magellanica* in such situations can be regarded as a disturbed seral community developed in response to bird and other animal activity. Small patches of *Deschampsia chapmanii*

Figure 6.5. Dense mixed *Stilbocarpa polaris–Poa foliosa* community on well-drained coastal slope, Sandy Bay. Scale marker total 80 cm.

with *Agrostis magellanica* and *Poa annua* occur in some areas. Often there is a subtle gradation from short grassland through mire to herbfield, the transition depending mainly on drainage patterns in the underlying peat.

On raised coastal terraces, herbfield covers large areas and grades into maritime tall tussock grassland. Complex communities of *Poa foliosa*, *Stilbocarpa polaris* and *Pleurophyllum hookeri* are widespread. Graminoids, other vascular plants, bryophytes and lichens occur in herbfields. Variations in vegetation composition between particular herbfield areas on raised coastal terraces may be due largely to water-table levels and drainage through underlying peats.

Pleurophyllum hookeri, a relatively large-leafed rosette plant, is a common herbfield plant (Figures 6.7, 9.6(a)). It occurs widely on raised

Figure 6.6. *Poa foliosa* lining a drainage stream across wet herbfield on raised coastal terrace, north of Bauer Bay.

coastal terraces in a variety of habitats. Morphology and vigour of plants and plant density are strongly influenced by environmental conditions. Increased waterlogging of the substrate generally results in a decrease in vigour as measured by rosette diameter. Soil oxygen flux correlates well with plant vigour, although nutrient availability is also likely to be important (Jenkin & Ashton, 1979).

Reproduction of *Pleurophyllum hookeri* is largely vegetative, from axillary shoots or rhizomes. It does not flower every year, but irregularly. In heavy-flowering seasons *Pleurophyllum*-dominated herbfield is a spectacular sight (Figure 7.2). In mire on the raised coastal terrace at Handspike Corner, *Pleurophyllum hookeri* occurs in long stripes (Figure 6.8).

Figure 6.7. *Pleurophyllum hookeri* in flower.

The origin of these stripes has not been ascertained but they are not aligned parallel to prevailing winds as believed by Jenkin & Ashton (1979).

Raised coastal terraces are crossed by streams from the plateau and by seepage-flush channels which drain water from the coastal slopes. Streams are often lined with *Poa foliosa*. Flush lines generally support heavy growths of *Juncus scheuchzerioides, Montia fontana* and the moss *Breutelia pendula*.

Extensive areas of bryophyte-dominated quaking mire, known on the island as 'featherbed', occur on the coastal terrace, particularly at Handspike Point and Half Moon Bay. The water table in such areas is at or very close to the vegetation surface and drainage is slow.

Figure 6.8. *Pleurophyllum* plants grow in slightly raised rows in a hepatic-dominated mire on raised coastal terrace, Handspike Corner.

Bryophyte-dominated communities occupy slightly depressed saturated channels between stripes of the *Pleurophyllum*.

At Handspike Corner, *Carex trifida* occurs in pure stands, associated with *Poa foliosa* along drainage lines, or with herbfield and short grassland species.

The raised coastal terrace at Handspike Corner is the only place on the island where *Poa litorosa* is common. Here, in association with *Festuca contracta*, it occupies slightly raised areas surrounded by bryophyte-dominated mire. Two small stands of the species also occur in upland short grassland communities near Caroline Cove at the south-western corner of the island.

Coastal slopes

The island plateau rises abruptly from fringing beaches or raised coastal terraces. The coastal slopes are generally covered by tall tussock grassland in which *Stilbocarpa polaris* often grows intermixed with tussocks of *Poa foliosa* (Figure 6.5). At different altitudes, depending on local conditions, *Poa foliosa* tussocks and *Stilbocarpa polaris* give way to short grassland dominated by *Luzula crinita* and *Agrostis magellanica* (and occasionally *Uncinia* spp.) (Figure 6.3).

As on coastal terraces, on slopes where almost pure stands of *Poa foliosa* form a closed canopy, there are few other vascular plants but some bryophytes such as *Brachythecium* spp., *Achrophyllum dentatum*, *Amblystegium serpens*, *Lophocolea* spp. and *Metzgeria* spp. do occur. Where canopy gaps have opened through tussock death, rabbit grazing, soil creep or landslip, the surface is colonised by herbs and bryophytes from the surrounding vegetation. *Agrostis magellanica*, *Acaena* spp., *Luzula crinita*, *Stilbocarpa polaris* and bryophytes such as *Hypnum*

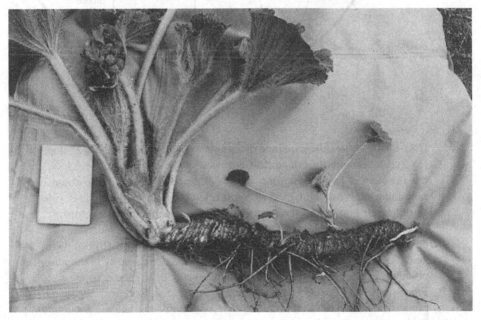

Figure 6.9. *Stilbocarpa polaris* has a fleshy rhizome from which adventitious shoots arise. This allows rapid establishment from rhizome fragments on disturbed sites, and rapid regrowth after severe rabbit grazing removes above-ground parts. Notebook (15 × 9 cm) gives scale.

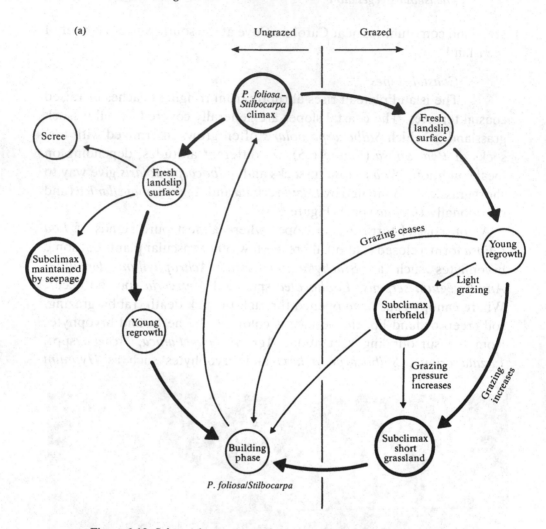

(a)

Ungrazed   Grazed

Figure 6.10. Schematic sequence of revegetation on disturbed coastal slopes. (a) Recolonisation of fresh landslip surfaces by *Poa annua* leads to development of *Poa foliosa/Stilbocarpa polaris* communities in the absence of rabbit grazing. With light grazing, a transitional subclimax herbfield develops. With heavier grazing, a subclimax short grassland community develops. Removal of grazing pressure from the subclimax communities leads to development of a *Poa foliosa/Stilbocarpa polaris* community. (b) Following heavy grazing of the *Poa foliosa/Stilbocarpa polaris* community in the absence of landslips, the bared ground is rapidly revegetated. In grazed areas, a subclimax short grassland develops and is maintained. Removal of grazing pressure results in development of a *Poa foliosa/Stilbocarpa polaris* community. Redrawn from Scott (1986).

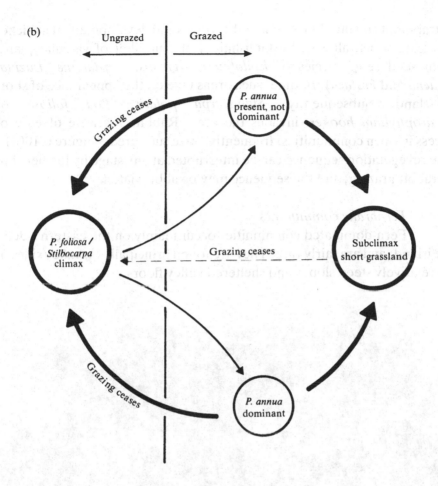

(b)

Ungrazed   Grazed

Grazing ceases

P. annua
present, not
dominant

P. foliosa /
Stilbocarpa
climax

Grazing ceases

Subclimax
short grassland

Grazing ceases

P. annua
dominant

cupressiforme, Breutelia pendula and *Tortula rubra* commonly occur in open tussock grassland on plateau slopes.

Very few plant species occur as understorey components of dense *Stilbocarpa polaris* stands on coastal slopes. *Stilbocarpa* produces large amounts of seed each year, but few seedlings are found in the dense shade under the complete canopy of leaves. Mice and rats harvest large quantities of its seed. Although it is often heavily grazed by rabbits, *Stilbocarpa* regenerates readily from adventitious buds along its rhizome (Figure 6.9).

Landslips are common on steep unstable slopes (Scott, 1985; Chapter 5) but are rapidly revegetated (Figure 6.10(a)). Bryophytes (species of *Marchantia, Jungermannia, Jamesoniella, Riccardia, Dicranella, Breutelia pendula, Ditrichum, Psilopilum, Polytrichum* and others) are early colonisers of shallow mineral soils. Establishment of bryophytes leads to entrapment and build-up of mineral particles and development of a thicker substrate which allows re-establishment of a number of vascular plants from seed (e.g. species of *Epilobium, Agrostis, Cardamine, Luzula, Acaena* and *Festuca*). In time, such areas take on the appearance of short grassland. Subsequently, *Stilbocarpa polaris, Poa foliosa* or *Pleurophyllum hookeri* invades the area. Rabbits, because of ease of access in open communities, frequently graze such areas (Figure 6.10(b)). The revegetation sequence can be interrupted at any stage by further slips or rabbit grazing, and the sequence may be abbreviated.

Fernbrake communities

Fern-dominated communities occur mainly on the eastern side of the island, and are fairly rare. They are found principally on valley sides or on relatively steep slopes and sheltered valley floors.

Figure 6.11. Dense *Polystichum vestitum*-dominated community in sheltered gully, Finch Creek.

Polystichum vestitum forms dense closed stands in relatively few localities (Figure 6.11). Few other plants occur beneath the canopy of fronds
which may be up to 70 cm high. More usually, the *Polystichum* occurs in
mixed communities with *Poa foliosa, Stilbocarpa polaris* and
Pleurophyllum hookeri. Taylor (1955a) considered *Polystichum* as an
associate or codominant of the *Poa foliosa–Stilbocarpa polaris* association
and, less frequently, the *Poa foliosa* association.

Blechnum penna-marina dominates fernbrake in only one locality
(Finch Creek) where it occurs amongst herbfield and short grassland. The
species is becoming more evident in many areas. This has been linked to
reduction in rabbit numbers in many parts of the island (Copson, 1984).

Two other ferns. *Grammitis poeppigeana* and *Hymenophyllum peltatum*, are relatively widespread, particularly in feldmark and other upland
localities, but are not important species for vegetation classification.
Lycopodium australianum is similarly widespread on the plateau but not
common.

Plateau uplands

A wide range of plant communities – tall tussock grassland, short
grassland, herbfield, mire and feldmark – occurs in upland habitats.
Exposure to prevailing westerlies, drainage and soil depth are important in
determining the composition of plant communities in upland areas.
Burrow-nesting petrels, mice and rabbits are widespread on the plateau.
Rabbit grazing, modified by the success of myxomatosis in dramatically
reducing rabbit numbers (Chapter 10), affects the relative abundance of
particular species.

Tall tussock grassland occurs in both sheltered and exposed upland
sites. Tussocks are generally smaller than those on coastal terraces and
slopes. It is possible that *Poa foliosa* may once have been more widely
distributed in upland areas. It is becoming more common amongst short
grassland where rabbit populations have been significantly reduced. Relict
tussock populations in comparatively inaccessible areas such as on rocky
outcrops and on islands in lakes lend support to this idea (Copson, 1984;
Scott, 1988).

Short grassland communities dominated by *Festuca contracta* are widespread but not extensive and grade imperceptibly into communities
dominated by *Agrostis magellanica* and *Luzula crinita* (Figure 6.12). In
many areas, particular vascular species have become more prominent in
short grassland communities since rabbit numbers have declined. *Uncinia
hookeri* and *U. divaricata*, both species grazed by rabbits, have shown a

Figure 6.12. Short grassland community dominated by *Agrostis magellanica* and *Luzula crinita* on plateau south of Green Gorge.

particularly noticeable resurgence. *Cerastium fontanum*, much favoured by rabbits, is spreading in all areas on the island from lowland to feldmark regions.

Three major types of upland herbfield can be distinguished. They are dominated by *Acaena* species, by *Pleurophyllum hookeri* and *Stilbocarpa polaris* and by *Azorella macquariensis* with *Pleurophyllum hookeri*.

Extensive *Pleurophyllum hookeri* herbfield covers large areas of the plateau in the northern third of the island (Figure 7.2). It is likely that *Pleurophyllum*-dominated herbfield was formerly far more extensive (Copson, 1984). Rabbits graze both the *Pleurophyllum* and *Stilbocarpa* heavily. In recent years, at least in the northern two-thirds of the island,

Figure 6.13. *Pleurophyllum hookeri–Azorella macquariensis* community on plateau south-west of Lake Ifould. Fault scarp bounds the south-eastern margin of lake. The lake drains via a north-easterly flowing creek from its northern end (arrow).

there has been a dramatic recovery in vitality of *Pleurophyllum* and spread of *Stilbocarpa*.

Pleurophyllum hookeri is associated with the cushion plant *Azorella macquariensis* in a number of upland sites, generally above 300 m (Figure 6.13). Death of a *Pleurophyllum* plant, or removal of its leaves by rabbits, may lead to destructive erosion of exposed *Azorella* cushions or carpets.

Mire communities may develop where the water table is high and drainage restricted. *Agrostis magellanica*, *Ranunculus biternatus* and bryophytes such as *Breutelia pendula*, *Bryum laevigatum* and *Riccardia*

Table 6.2. *Feldmark vegetation on Macquarie Island. Percentage occurrence at 2000 points of first contact along 20 m line transects.*

	Location of site				
	Near Boot Hill, 225 m asl	N of Prion Lake, 175 m asl	Windy Ridge, 325 m asl	SE Mt Gwynn, 300 m asl	E side Mt Law, 300 m asl
Vascular plants	7	1	9	33	65
Bryophytes	16	23	69	16	24
Lichens	0	2	7	5	1
Total vegetation	23	26	85	54	90
Selected species:					
Azorella macquariensis	5	1	8	30	50
Ditrichum strictum	15	17	14	1	0
Rhacomitrium crispulum	0	1	18	11	13
Andreaea acuminata	0	0	23	0	0
Total no. of species	8	14	20	25	23
Number of transects	3	2	2	2	1

Data from Selkirk & Seppelt (1984).

cochleata are common. *Deschampsia caespitosa* is generally confined to such habitats but is uncommon. One moss species, *Trematodon flexipes*, is often found on sodden bare peats in seepage mire habitats where it has no competition from other plants.

Mires dominated by *Juncus scheuchzerioides* are most common in lowland areas but also occur on plateau uplands. *Montia fontana, Ranunculus biternatus, Colobanthus quitensis* and *Cardamine corymbosa* are commonly associated with the *Juncus*, and bryophytes such as *Breutelia pendula* and *Drepanocladus aduncus*.

Feldmark occupies much of the windswept higher part of the plateau. Feldmark is a complex vegetation formation. It is 'an open subglacial community of dwarf flowering plants, mosses and lichens' (Beadle & Costin, 1952). At high altitude on Macquarie Island, hepatics are also important in binding soil and gravel.

Vegetative cover in feldmark is extremely variable and large expanses of gravel-covered ground bare of vegetation are common (Table 6.2). The cushion-forming *Azorella macquariensis* is the dominant vascular plant but much of the Macquarie Island feldmark is bryophyte-dominated. Polsters of *Ditrichum strictum* (Figure 6.14) and tufts or turves of *Rhacomitrium crispulum* (Figure 6.15) contribute significantly to productivity

Table 6.3. *Standing crop and production of mossfield and feldmark on Macquarie Island. (Values corrected to 100% cover in parentheses.)*

	Standing crop (g m^{-2})	Cover (%)
Feldmark[a]	193[c] (495)	39 (100)
Mossfield[b]	145[c] (659)	22 (100)

[a] 61.5% *Azorella macquariensis*, 36.6% *Rhacomitrium crispulum*, assorted herbs 1.9%.
[b] 99% *Ditrichum strictum*, 1% *Agrostis magellanica*, *Luzula crinita*, *Ranunculus biternatus*, *Rhacomitrium crispulum*, *Andreaea acutifolia*.
[c] Green shoot biomass above peat.

Data from Seppelt & Ashton (1978).

and biomass of feldmark (Table 6.3) (Seppelt & Ashton, 1978).

Other mosses (*Andreaea* spp., *Rhacocarpus purpurascens*, *Rhacomitrium lanuginosum*), hepatics (particularly *Jamesoniella colorata*), lichens (species of *Pertusaria*, *Lepraria*, *Stereocaulon*) and small vascular plants (*Agrostis magellanica*, *Luzula crinita*, *Ranunculus biternatus*) are generally only minor components of the total vegetative cover.

Figure 6.14. *Ditrichum strictum* polsters and cushions of *Azorella macquariensis* in feldmark, Boot Hill.

Figure 6.15. *Rhacomitrium crispulum* forms rows aligned almost parallel to prevailing wind direction in feldmark on an exposed ridge, south-east of Mt Fletcher, adjacent to 'Lake Barker'.

Taylor (1955a) recognised two alliances within the feldmark formation: an *Azorella selago* (now *A. macquariensis*) alliance and a *Dicranoweisia antarctica* alliance. Ashton & Gill (1965) realised that Taylor's *Dicranoweisia* was, in fact, *Ditrichum strictum*. While *D. strictum* is abundant in many feldmark localities, *Rhacomitrium crispulum* is more abundant generally in plateau uplands and feldmark. Ashton & Gill (1965), while acknowledging the complexity of the feldmark community, considered it as a single vegetation unit, a complex *Azorella – Ditrichum* alliance.

There are many areas transitional between typical feldmark and typical short grassland communities. All the vascular plant species, bryophytes, and most of the lichen species of short grassland and plateau herbfield associations are found as minor components in feldmark communities.

Rocky outcrops in feldmark provide additional niches for colonisation by all plant groups. The filmy fern (*Hymenophyllum peltatum*) was first found in such localities and, although not restricted to such plateau outcrops and higher peaks, is most commonly found there. Bryophytes such as *Macromitrium longirostre, Rhacocarpus purpurascens, Frullania rostrata* and *Plagiochila retrospectans*, and lichens such as *Stereocaulon ramulosum* and many crustose species, are abundant on rocky outcrops.

Vegetation patterns in feldmark range from regularly alternating stripes of vegetation and gravel to irregular vegetation patches amongst gravel. Regular striping may be associated with steps and stairs terrain. Leeward terraces on east-facing slopes, on which vegetation forms the risers and gravel forms the treads of the stairs, were described by Taylor (1955b). Windward terraces on west-facing slopes are less regular, with soil and gravel risers and vegetated treads. Taylor interpreted both terrace types as responses by vegetation to the present prevailing strong westerly winds. Löffler, Sullivan & Gillison (1983) and Löffler (1983) regarded windward terraces as wind formed, but leeward terraces as relict solifluction land-forms developed during a former more severe climate. They described terraces on slopes with neither windward nor leeward aspects as inter-mediate in form. Ashton & Gill (1965) described in detail the structure of terraces on windward slopes on North Mountain and the involvement of wind in their formation, maintenance and gradual upslope progression.

From an examination of a wide variety of terraces on non-windward slopes, Selkirk, Adamson & Seppelt (1988) concluded that wind, moist-ure, hillslope angle, slope stability and vegetation all affect terrace form. They observed dynamic features of terraces which show that they are not merely inherited landforms from a former glacial period.

They found vegetation disposition in feldmark to be clearly associated with hillslope angle. On gently sloping ground (slopes of about 0–15°), hillsides are markedly terraced approximately along the contours, with vegetated terrace risers and gravel-covered terrace treads (Figures 6.16, 6.17(a)). Treads slope slightly (1°–2°) either outwards or inwards, and usually slope slightly (0°–5°) along their length. These are the leeward terraces of Taylor (1955a,b), Löffler, Sullivan & Gillison (1983) and Löffler (1983).

On well-developed terraces, there is a marked difference in vegetative cover and composition between risers and treads. *Azorella macquariensis* is a major component of risers, forming almost complete cover in some areas. More usually it forms mixed associations with short grassland species, bryophytes, lichens and, occasionally, *Pleurophyllum hookeri*.

Figure 6.16. Terraces in feldmark on the east-facing slope of Mt Martin. Terrace treads approximately 1.5 m wide have *Ditrichum strictum* scattered on gravel. Terrace risers, 2–4 m wide, have continuous cover of *Azorella macquariensis*, smaller flowering plants and mosses.

Terrace treads are generally covered with stones, gravel and fines with open vegetation dominated by mosses, particularly *Ditrichum strictum*, *Rhacomitrium crispulum* and *Andreaea* spp., with scattered hepatics and vascular plants.

Terrace and vegetation dynamics yield several minor variations in the boundaries between plants and gravel. At the back of a tread, near the base of the vegetation of the riser above, stirring by needle ice action followed by water washing down the riser can gradually erode fine material away (Figure 6.17(b)). Over time, this will steepen the angle of the riser, and, as vegetation slowly grows into gravel at the front of the tread above, the riser will move upslope. In other cases the back of the tread may be well-stabilised by vegetation extending onto the tread from the riser above (Figure 6.17(c)).

Frost sorting of surface gravel on the treads is common, with larger stones sorting to the front, or outer edge of the tread. Similar sorting occurs on the chutes which lead from one tread to the tread below. Stone stripes are common on gravel treads. Small sorted polygons occur on terrace treads at a few sites (Chapter 5).

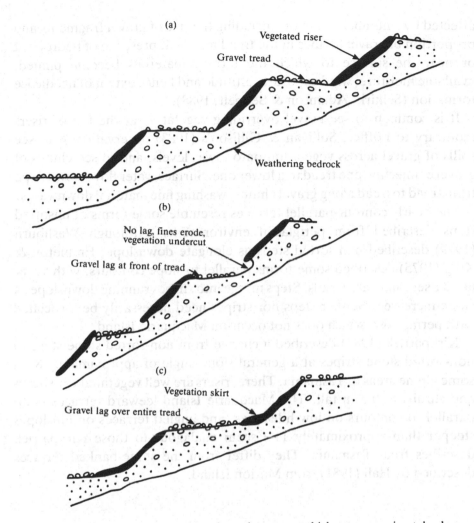

Figure 6.17. Notional sections through terraces which run approximately along contours. (a) Risers are vegetated, treads have gravel lag. (b) Vegetation on risers can be undercut at base when fines are eroded from a surface lacking gravel lag. (c) Vegetation extending as a skirt from the riser above can help stabilise the back of a tread. From Selkirk, Adamson & Seppelt (1988), reproduced courtesy Royal Society of Tasmania.

In a few places completely vegetated terraces occur: the surface of the terrace tread has become covered with vegetation, presumably having stabilised after fines have been washed out, frost heaving has decreased, and additional supply of debris from above has been exhausted.

On more steeply sloping sites, both terraces and associated vegetation and gravel stripes are oriented at increasingly steep angles to the contours. This change from gently to steeply sloping terraces occurs at a hillslope angle of approximately 11°. At a particular site this threshold seems to be affected by a number of factors including the size of gravel fragments and proportion of gravel to fines in the tread material, presence of bedrock at or near the surface to which the terrace materials become pinned, available moisture, wind exposure, altitude and hence extent of needle ice formation (Selkirk, Adamson & Seppelt, 1988).

It is common to see gravel overriding vegetation at the top of risers (contrary to Löffler, Sullivan & Gillison's (1983) observations), to see spills of gravel across vegetation onto lower levels, and to see chutes of gravel connecting one tread to a lower one. Surface water flows downslope from tread to tread along gravel chutes, washing fine material downwards.

The nearly contour-parallel terraces resemble some forms of unsorted steps described from periglacial environments. Although Washburn (1979) described non-sorted steps as elongate downslope, Embleton & King (1975) described some forms parallel to slope contours, with vegetated risers and bare treads. Steps merge into stripes running downslope as slope increases. Neither steps nor stripes need necessarily be associated with permafrost, which does not occur on Macquarie Island.

Kirkpatrick (1984) described a change from non-sorted stone steps to non-sorted stone stripes at a general slope angle of approximately 8° in some alpine areas of Tasmania. There, risers are well vegetated with herbs and shrubs to 0.5 m tall. The Macquarie Island leeward terraces (sub-parallel to contours on low hillslopes, and angled terraces on hillslopes steeper than approximately 11°) seem analogous to those Kirkpatrick describes from Tasmania. They differ from the stone-banked terraces described by Hall (1981) from Marion Island.

Influence of animals on vegetation

Rabbits have had a marked effect on island vegetation. Early reports indicate that *Pleurophyllum hookeri* and *Poa foliosa* were formerly much more widespread and that grazing by rabbits has been largely responsible for the change to present conditions. There is evidence to suggest that, with myxoma virus control of rabbits, populations of both plants are increasing. Other plants such as *Uncinia* species and *Cerastium fontanum* are rapidly becoming more prominent as grazing pressure decreases. Rabbits have played a significant role in the spread of *Acaena* species over the island. *Acaena* fruit, with its hooks, is ideally adapted for spread by animals (Bergstrom, 1986). Although rats and mice eat seeds and other herbage there is no evidence to suggest they have any impact on the vegetation.

Royal and king penguins denude some areas of coastal and inland vegetation and remove underlying peat in establishing permanent colonies. High nutrient levels in the vicinity of such colonies may have beneficial effects on at least some species, such as *Poa cookii*. Gentoo penguins induce local change in vegetation by destroying tall tussock grassland and herbfield, but as they move their colonies each year, their effects are not catastrophic (Figure 8.2).

Southern giant petrels nest colonially and when the same nesting territories are used for several years the composition of the vegetation on and around the site may change significantly.

Coastal communities of *Poa foliosa* are favoured haul-out sites for elephant seals. As a result, many tussocks are destroyed or damaged each year as the seals lie on them, rub against the pedestals, defecate and urinate. In some such sites, killed tussocks are replaced by *Poa annua, Cotula plumosa* and, in wetter sites, *Callitriche antarctica* (Chapter 9).

Gillham (1961) described the effects of birds and elephant seals on the island's vegetation. Smith (1976, 1979) described the influence of seabirds on the nutrient status of soils on Marion Island. He found both nitrogen and phosphorus levels were considerably higher near bird nests. Brothers (1984) implied that, on Macquarie Island, nutrient input from burrow-nesting petrels influenced at least the colour of *Poa foliosa* leaves near nest sites but no quantitative studies have yet been made.

Introduced and alien plant species

Vascular plants introduced, or presumed to have been introduced, to the island have been listed by Jenkin, Johnstone & Copson (1982) (see Table 12.2; Appendix 1). Among the species listed, at least *Corybas macranthus* is not introduced. First reported in 1977 (Brown, Jenkin, Brothers & Copson, 1978) from a limited area on the north-west coast raised beach terrace, it occurs in a number of widely separated localities on the island.

Rumex crispus is known from only one locality, north of Bauer Bay. First seen in 1980 (Copson & Leaman, 1981), the species is now represented on the island by two plants, implying either successful seed-set in one year since its discovery or the possibility of at least two separate seeds being introduced. Like *Corybas*, its mode and time of arrival on the island are unknown.

Galium antarcticum is similarly known from only one locality, near Skua Lake (Seppelt, Copson & Brown, 1984). It is a small plant and one which could easily be overlooked: it may be more widespread on the island. Its affinities with New Zealand species have yet to be evaluated.

Cerastium fontanum and *Stellaria media* were first reported by Hamilton (1895) and may have been introduced as a result of human occupation. *Poa annua* was first reported in 1880 (Scott, 1883). The species is presumed to have been introduced to Macquarie Island by humans but there is evidence for its existence on Iles Kerguelen before European discovery of that island group (Bellair & Delibrias, 1967). *Poa annua* has now been found at two sites, both on recently exposed glacial moraines, on Heard Island (Scott, 1989) but it is not clear whether it has been introduced by humans or has reached Heard Island independently.

It is of interest that populations of *Poa annua* on Macquarie Island are genetically distinct from other populations of the species (Ellis, Lee & Calder, 1971; Calder, 1973).

A number of agricultural or weedy species has been deliberately or accidentally introduced to Macquarie Island. These have either died out or been removed (Jenkin, Johnstone & Copson, 1982). Vegetable and horticultural seeds have been planted on the island over many years with (thankfully) little success except in heated glasshouses (Chapter 12).

The low temperatures, particularly during the summer growing season, poor soil, and high deposition of air-borne salt are important factors limiting the potential for survival of deliberately- or naturally-introduced species. Only those species ecologically adapted to the rigours of a subantarctic environment have been, or will be, able to survive.

Only one bryophyte species, *Funaria hygrometrica*, could be considered as a possible introduction by humans (Chapter 12). There is no available evidence of introduced fungi. Fungi in the subantarctic are likely to be cosmopolitan species spread by wind. The world's fungal flora as a whole is still relatively poorly known except for parasites of agricultural plants and common saprophytes.

Biogeographical relationships of the flora

Wind, water and birds have all played a part in the dispersal of organisms to subantarctic islands (Bergstrom & Selkirk, 1987). Few plants, plant parts, seeds or propagules could survive the length of time immersed in or exposed to sea-water necessary for passive flotation as a means of dispersal to most of the islands, although some moss propagules show considerable ability to survive immersion in seawater. In the westerly drift system, pumice, presumably from the South Sandwich Islands eruption in March 1962, reached the shores of Macquarie Island, Australia and New Zealand some two years later. Objects travel passively in the ocean currents at speeds of about 13 km per day or up to 30 km per day for larger fragments which may be wind-assisted (Deacon, 1960; Longton, 1977).

Long-distance dispersal by winds, both jet streams and lower-level systems, and by birds as agents of dispersal, seem the most likely and only plausible explanations for the origin of the floras and faunas of the subantarctic islands with the exception of deliberate or accidental intro-ductions by humans.

Seeds of *Acaena* and *Uncinia* species bear hooked spines, making them well-adapted for transport by animals. Other seeds, such as those of *Cotula plumosa*, have sticky seed coats. Some fleshy seeds, such as those of *Coprosma pumila*, may be ingested by birds. Seeds of some species, such as *Cotula plumosa, Pleurophyllum hookeri* and *Epilobium* spp., have a feathery or hairy pappus (Bergstrom, 1986) which could aid in wind dispersal although field observations on Macquarie Island indicate that, perhaps because of the damp climate, seeds of these species are dispersed only short distances from the parent plant. Very small seeds, such as those of *Corybas macranthus*, and small spores and other propagules of bryo-phytes, fungi, ferns and algal cells are ideally suited for dispersal by wind.

The diversity of the terrestrial faunas and floras of subantarctic islands is limited largely by their geographical isolation, small size, relatively small areas suitable for colonisation and low temperatures of the summer growing season. Smith (1984) listed 24 species of grasses, 32 forb species and 16 pteridophyte species, together with 250+ mosses, 300+ lichens and about 70 macrofungi from subantarctic islands. Taxonomic relationships of many of these species are still unclear. Early accounts of the vegetation of subantarctic islands listed many species as endemic. Recent taxonomic revisions are proving this to be generally not so. As Walton (1986: 297) states:

> Good phytogeographical accounts must wait on good taxonomy and until this is forthcoming critical discussion of biogeographical patterns must remain tentative.

Taxonomic difficulties still prevent any reasonable biogeographic dis-cussion of the lower plants. Amongst lichens and bryophytes there appears to be a high proportion of endemic species and, with taxonomic revisions, a significant bipolar element.

As the subantarctic islands lie in the general westerly wind and current drift systems it is not surprising that their floristic affinities are predomi-

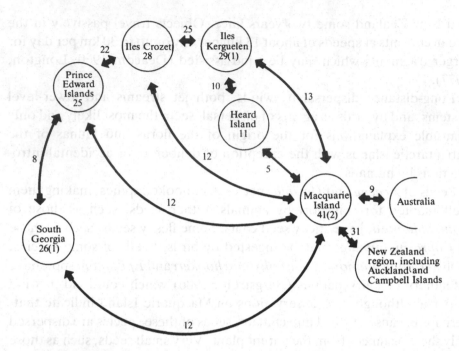

Figure 6.18. Biogeographic relationships of the indigenous vascular flora of the subantarctic islands. For each island or area, numbers indicate indigenous species, numbers in parentheses indicate endemic species. Numbers on arrows indicate species common between areas. These figures exclude *Poa annua* as its status is equivocal.

nantly generally westerly (Figure 6.18). The South Georgian flora possesses a strong Fuegian element. This diminishes in an easterly direction through all the subantarctic islands. The floras of Marion and Prince Edward Islands and Iles Crozet have a limited southern African affinity while that of Macquarie Island shows a strong Australasian relationship. There is also a uniquely subantarctic element which is found in all groups of the subantarctic flora. The most widely distributed introduced plants are *Cerastium fontanum, Stellaria media* and *Poa annua*, the cosmopolitan *Poa annua* possibly being indigenous. *Poa annua*, normally an annual, behaves as a perennial in the subantarctic and is ecologically important on Macquarie Island as a coloniser of disturbed habitats. Similarly, on South Georgia, *Poa annua* colonises areas of natural vegetation damaged by grazing reindeer (Walton, 1986).

7

Lakes

Numerous lakes, ponds, tarns and pools dot the surface of Macquarie Island. The almost constant precipitation and low rate of evaporation ensure a plentiful supply of water. Surface peats are waterlogged much of the time, mires are extensive, creeks are numerous.

Open bodies of freshwater on Macquarie Island can be grouped into three categories:

(a) ponds in peats on the coastal terrace (Figure 7.1),

(b) ponds in peat beds on the plateau (Figure 7.2),

(c) lakes and ponds in tundra soils and rocky substrates on the plateau (Figures 3.5, 6.15) (Evans, 1970).

Ponds with peat substrates on both the coastal terrace and plateau are formed by freshwater filling irregularities and depressions in the surface of the waterlogged peat. In line with the various views of origin of landforms on the island, ideas on the origin of lakes vary. Some authors have suggested that lakes and ponds on tundra soils, rocky substrates and in rocky valleys may be of glacial origin, formed in ice-deepened valleys or in moraine-dammed depressions (Evans, 1970). Recently it has become clear that many lakes are fault-controlled, occurring in lineament-bounded depressions, or are formed by fault-damming of watercourses (Ledingham & Peterson, 1984) (Chapter 5).

There are twenty named lakes of medium size on the plateau (Figure 7.3) and six named lagoons and tarns on the island. The term 'tarn' is used on the island to mean 'small lake' or 'large pond'.

Only Major Lake, the largest on the island (Figure 3.5), Tiobunga Lake, Prion Lake and Floating Island Lake (Figures 7.5, 7.6) have been studied in any detail with respect to form and depth (Evans, 1970; Peterson, 1975; Gardner, 1977). Maximum depths of several other lakes are known (Peterson, 1987). Major and Tiobunga Lakes are both quite shallow (Table 7.1). Peterson (1975) regarded a glacial origin for Major and

Figure 7.1. Sheltered pond in mire, raised coastal terrace, Mawson Point.

Tiobunga Lakes as most unlikely. He regarded Prion Lake with its steep sides on the east, west and northern shores, and shallow southern end, as a likely candidate for a glacially-formed lake. He suggested that Mt Eitel could have acted as a 'snow fence' in the path of the westerly winds and concluded that the Prion Lake drainage basin offered the most favourably-oriented site on the island for lee-side ice accumulation. Subsequently, Ledingham & Peterson (1984) concluded that Prion Lake, in common with Tulloch Lake and Lake Ifould, lies in a fault-bounded depression rather than in a glacially scoured basin (Figure 5.1). Scoble Lake, Square Lake, Flynn Lake, Lake Gratitude and Lake Ainsworth (Figure 3.2) are also fault-dammed lakes (Chapter 5).

Figure 7.2. Plateau pond, south-west of North Mountain (the 'S' pond of Evans, 1970), in *Pleurophyllum hookeri* herbfield. Encroaching vegetation is *Ranunculus biternatus*.

Chemistry of Macquarie Island lake waters

Macquarie Island's lake, tarn and pond waters receive most of their ionic constituents from atmospheric precipitation with little contribution from surrounding rocks (Evans, 1970; Tyler, 1972; Buckney & Tyler, 1974).

Anderson's (1941) oceanic origin factor (OOF) gives a measure of the proportion of major ions in inland waters that are of marine origin:

$$OOF\ (\%) = \frac{(Cl\ m.\ eq.\ l^{-1}) \times 1.107 \times 100}{(total\ anions\ m\ eq^{-1})}.$$

Figure 7.3. Location of principal lakes, tarns and watercourses. Sc = Scoble
Lake, Is = Island Lake, Em = Emerald Lake, DL = Duck Lagoon, Sq =
Square Lake, Tu = Tulloch Lake, Su = Surry Lake, LP = Little Prion Lake,
Pr = Prion Lake, MT = Midas Tarn, Ei = Lake Eitel, Co = Concord Lake, If
= Lake Ifould, BT = Betsey Tarn, Sk = Skua Lake, ST = Skua Tarn, GT =
Green Gorge Tarn, Gr = Gratitude Lake, CT = Cumberland Tarn, Fl = Flynn
Lake, Py = Pyramid Lake, En = Endeavour Lake, Ma = Major Lake, Ti =

Table 7.1. *Morphological characteristics of Macquarie Island lakes.*

Lake	Area (km²)	Max. depth (m)	Volume (m³ × 10⁶)	Max. length (km)	Mean length (km)
Major[a]	0.5	16.2	2.37	0.97	0.46
Tiobunga[a]	0.07	2.3	0.10	0.45	0.16
Prion[b]	0.3	32.3	6.35	1.10	0.29
Square[b]	0.09	0.6	0.03	0.38	0.24
Floating Island[c]	0.008	6.9		0.12	
Skua[d]		1.5			
Gratitude[d]		17.1			
Flynn[d]		19.8			
Waterfall[d]		25.2			

[a] Peterson (1975).
[b] Evans (1970).
[c] Gardner (1977).
[d] Peterson (1987).

High OOF values for both lakes and rainwater collected on the island (Table 7.2) reflect the fact that prevailing strong westerly to north-westerly winds are heavily laden with salt from the ocean. Sea-salt becomes airborne when air bubbles, trapped at wave crests, burst, releasing seawater droplets which evaporate to leave small salt particles in the atmosphere light enough to be carried upwards by air currents. Airborne sea-salt 'scavenged' from the atmosphere forms the major component of salts of marine origin in precipitation falling on Macquarie Island. The minor component comes from airborne sea-spray droplets trapped by precipitation before evaporation has left solid salt particles (Mallis, 1985, 1988). The rugged nature of the island's west coast, with its many rocks and reefs over which waves break, results in airborne sea-spray becoming incorporated into winds blowing over the island.

Concentration of salts in lakes is greatest close to the west coast, and decreases generally eastwards across the island (Evans, 1970; Tyler, 1972). Although Tyler (1972) ascribes this to most abundant deposition of wind-borne sea-spray closest to the western seaboard of the island, Mallis (1988) very plausibly accounts for the gradient in terms of sea-salt

Tiobunga Lake, Wa = Waterfall Lake, Ai = Lake Ainsworth, NC = Nuggets Creek, RC = Rookery Creek, StC = Stony Creek, FC = Finch Creek, FlC = Flat Creek, RR = Red River, SwC = Sawyer Creek.

Table 7.2. *Oceanic origin factor (OOF) calculated for Macquarie Island freshwater bodies.*

100% > OOF > 91%	90% > OOF > 81%	80% > OOF > 71%	70% > OOF > 61%
Emerald[a] (2)			
Tulloch (2)	Tulloch (3)		
Prion (1, 2)	Prion (3)		
	Little Prion (2)		
Surrey[b] (2)			
Pyramid (2)			
Major (2)	Major (3)		
Tiobunga (2, 3)			
Duck Lagoon (1)			
6 coastal terrace ponds (1)			
2 North head ponds (1)			
Plateau pond (1)			
Scoble (1)	Scoble (2)	Scoble (3)	
Precipitation, Isthmus (1)	Rain, Isthmus (3)		
Precipitation, Prion Lake (1)			
	Island (2)		
	Flynn (2, 3)		
	Waterfall (2)	Waterfall (3)	
	Betsey Tarn (3)		
	Pyramid (3)		
	Cumberland Tarn (3)		
		Square[c] (1, 2)	Square (3)
		Green Gorge Tarn (2, 3)	
		Midas Tarn (3)	
		Concord (3)	
		Eitel (3)	
		Skua Tarn (3)	
			Skua (3)
		Gratitude (3)	
			Ifould (3)

Mean OOF of Macquarie Island lakes = 92.2% (1)
= 89.3% (2)
= 80.0% (3)

Values of (3) are lower than those of (2) probably because of prolonged storage of samples prior to analysis.

[a] As 'Bauer Lake'.
[b] As 'East Prion Lake'.
[c] As '23U Pond'.

Data from (1) Evans (1970); (2) Tyler (1972); (3) Buckney & Tyler (1974).

scavenged from the air by settling mist. Prevailing westerly winds approaching the island are moisture- and salt-laden. As the air-mass rises over the island it cools, forming a lenticular cloud (Figure 3.11). Moisture droplets from this cloud, settling as mist on the plateau and into lakes, take with them sea-salt scavenged from the air. Sites on the eastern side of the plateau receive less salt from the mist than western sites since scavenging has been occurring as the cloud moves eastwards across the island, leaving behind an air-mass progressively depleted of suspended dry sea-salt particles.

Mallis (1988) calculated that precipitation deposits some 2423 kg ha^{-1} yr^{-1} NaCl on the isthmus (6 m asl), 655 kg ha^{-1} yr^{-1} NaCl on the top of Wireless Hill (100 m asl) and 370 kg ha^{-1} yr^{-1} NaCl at 220 m asl at the north-east corner of the plateau. On the eastern coastal plain of Marion Island, sodium chloride in precipitation has been calculated at 350 kg ha^{-1} yr^{-1} in 1972-73 and at 267 kg ha^{-1} yr^{-1} in 1979-80 (Grobbelaar, 1978; Smith, 1987b).

The considerably lower sodium chloride content of precipitation measured at the Marion Island site, compared with the Macquarie Island isthmus site, both close to sea-level, may simply be a reflection of the relative shelter of the Marion Island site from the prevailing westerly to north-westerly winds.

Although precipitation is the major source of ions in the bodies of freshwater on Macquarie Island, some modification of proportions of dissolved salts occurs when water comes into contact with a lake's substrate. Water of a lake formed in a deep peat bed or a chemically inert rock basin receives very little dissolved mineral material from the substrate. Such lakes on Macquarie Island have waters with a high OOF and the same order of ionic dominance as seawater: Na > Mg > Ca > K : Cl > SO$_4$ > HCO$_3$. Such water bodies include the coastal terrace ponds and Duck Lagoon which Evans (1970) described in his category (a); the North Head ponds and plateau pond 'S' described in his category (b) and some category (c) lakes with rocky substrates, such as Tulloch, Prion and Pyramid Lakes (Table 7.2).

A lower OOF and an order of ionic dominance different from that of seawater indicates inclusion of minerals dissolved from the substrate. The water of Square Lake, with an ionic dominance Na > Ca > Mg > K : Cl > HCO$_3$ > SO$_4$ shows the greatest geochemical influence. The water in both Waterfall and Skua Lakes and Skua Tarn has a slightly lower, but still appreciable, geochemical influence (Tyler, 1972; Buckney & Tyler, 1974). The higher levels of CaCO$_3$ in these lakes is probably due to contact of water with lenses of interstitial calcareous material in pillow lavas on which the lakes occur. Analysis of interstitial *Globigerina* ooze from North Head shows a 49% calcium carbonate content (Mawson, 1943).

Bodies of freshwater on Campbell Island, like those on Macquarie Island, receive most of their solutes from seawater (mean OOF = 91.3% (Taylor, 1974a)). Lake waters on Iles Kerguelen and Auckland Island are more modified by island geochemistry, although still receiving substantial

marine input. Samples of lake waters from Iles Kerguelen have a mean
OOF of 72.4% (Evans, 1970; Tyler 1972); samples from Auckland Island
have a mean OOF of 73.8% (Taylor, 1974b).

With total dissolved solids (TDS) less than 173 mg per litre and calcium
ion concentration less than 0.75 milliequivalents per litre, Macquarie
Island freshwaters are generally soft by world standards. They are similar
to bogwaters in the Falkland Islands (Tyler, 1972).

Most bodies of freshwater on Macquarie Island are close to neutral pH.
Square Lake is slightly alkaline due to its high bicarbonate ion concen-
tration (Tyler, 1972). Brothers Lake is also alkaline (Hughes, 1986).
Measurements of pH from other lakes range from 5.7 to 7.2 (Tyler, 1972;
Hughes, 1986; Peterson, 1987).

Plateau lakes in rocky basins (Evans' category (c) lakes) receiving run-
off via well-drained tundra soils have very clear water. Some ponds and
pools on the coastal terrace and on peat bases (Evans' categories (a) and
(b)) have a little humic coloration.

Lake water temperatures

Evans (1970) studied water temperatures in four lakes over the
course of 15 months (Figure 7.4). Temperatures of different freshwater
bodies on the island depend on altitude, mean depth and volume. The
large volume of water in Prion Lake (Table 7.1) is kept in circulation by a
combination of continuous wind across the surface and a slow rate of
warming in summer (Figure 7.4). No thermal stratification occurred. Daily
temperature variation never exceeded half a degree. Ice did not form in
winter.

In the smaller water bodies he studied, Evans (1970) found daily
temperature variations to be much greater (up to 6 degrees in plateau pond
'S'). Water temperatures in pond 'S' (Figure 7.2) and in shallow Square
Lake (Figure 5.6) reflected ambient air temperatures quite closely, especially during winter months when hours of sunshine are particularly
short. During summer, when incident radiation is considerably higher
(Table 3.1), all freshwater bodies studied became warmer than ambient
temperature (Figure 7.4).

No lake on the island froze completely during Evans' (1970) study
period, although complete surface ice cover up to 10 cm thick occasionally
formed on small ponds. Such a thin surface ice cover over a lake in 1948 led
to the death of Charles Scoble and the naming of the lake after him
(Chapter 2).

Taylor (1955a) may have experienced more severe temperature con-

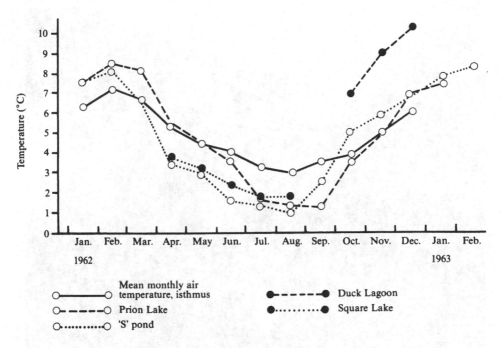

Figure 7.4. Mean water column temperatures for various water bodies.
Redrawn from Evans (1970).

ditions in 1950–51 than did Evans in 1965. Taylor described ice 18 inches
(45 cm) thick near lake shores, though only one inch thick (2.5 cm) over
the centre, and some small pools frozen solid.

Lake vegetation

Chemical and temperature characteristics of lake water are important in determining the distribution of freshwater animals and plants.
Plateau lakes such as Prion Lake are oligotrophic – having low nutrient
levels – and support relatively small populations of planktonic plants and
animals.

The aquatic angiosperm *Myriophyllum triphyllum* forms dense stands in
various shallow lakes with silty bottoms (Square Lake, Brothers Lake,
Scoble Lake, Green Gorge Tarn and Duck Lagoon). In the somewhat
deeper Floating Island Lake, *Myriophyllum triphyllum* grows extensively
around the margins where the water is less than 2 m deep (Figure 7.5). Two
islands, one 35 × 12 m, the other 30 × 30 m, float on the lake and are blown
from place to place on its surface (Figure 7.6). The islands are up to 2 m
thick, composed entirely of vegetation.

Gardner (1977) suggested that these islands may have originated by

Figure 7.5. Floating Island Lake with dense emergent *Myriophyllum triphyllum* and encroaching *Juncus scheucheroides, Agrostis magellanica* and *Poa annua* at edge.

detachment from the lake margin or lake bottom of large masses of the *Myriophyllum* which subsequently became colonised by a variety of herbfield and grassland species. Probing in 1987 failed to reveal a *Myriophyllum* basis to the islands. An alternative hypothesis for the origin of the floating islands involves vegetation sliding gently during a landslip from the western shore (arrowed, Figure 7.6), remaining as a coherent mat and floating on the lake surface, subsequently becoming colonised by additional plants.

Exposed in the cliff face half a kilometre to the north of Floating Island Lake is a lacustrine deposit which is a remnant of Palaeolake Nuggets. Like Floating Island Lake, Palaeolake Nuggets was perched on the rim of

Figure 7.6. Floating Island Lake from the north-west. The two islands are detached, and are blown about on the lake. The arrow shows an area on the western shore from which vegetation forming the islands may have slipped.

the Rookery Creek – Nuggets Creek basin. Exposed in the cliff face at the western margin of the plateau near Skua Lake, lacustrine deposits provide evidence for a former, large Palaeolake Skua (Chapter 5).

Although not collected before 1950, and then not observed below 300 feet (= 85 m) asl (Taylor, 1955a), *Myriophyllum* sp. has been part of the Macquarie Island flora throughout the Holocene and terminal Pleistocene. Well-preserved *Myriophyllum* sp. occurs in sediments dated at 9400 ± 220 radiocarbon years BP (SUA-1894) from Palaeolake Nuggets and $12\,470 \pm 140$ radiocarbon years BP (Beta 20165) from Palaeolake Skua (Selkirk, Selkirk, Bergstrom & Adamson, 1988).

Some *Ranunculus biternatus* and *Callitriche antarctica* occur with

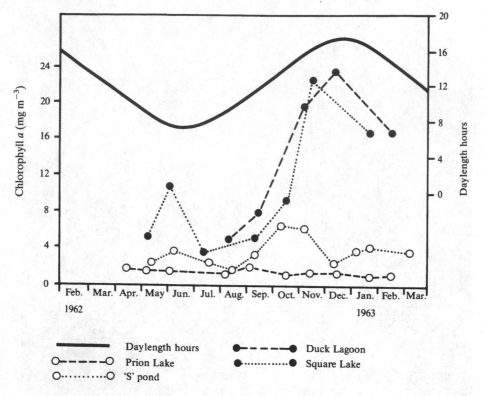

Figure 7.7. Seasonal distribution of chlorophyll *a* in four water bodies. Redrawn from Evans (1970).

Myriophyllum triphyllum in Square Lake (Hughes, 1986). *Callitriche antarctica* grows abundantly around the edges of, and on the surface of, old elephant seal wallow sites, now abandoned or rarely used, and on the edges of coastal pools enriched with seal excreta.

Several bryophytes (e.g. *Fissidens rigidulus, Tridontium tasmanicum, Ditrichum strictum, Pachyglossa tenacifolia* and *Drepanocladus aduncus*) occur in various plateau lakes (including Prion and Scoble Lakes), plateau ponds and in Green Gorge Tarn. These and other bryophytes are abundant in creeks and waterfalls on the island.

Planktonic green algae and diatoms are found in the lakes, ponds and creeks. Algae also grow as epiphytes on larger aquatic plants and terrestrial flowering plants. The island's freshwater algae have yet to be extensively studied, but include cyanophytes, desmids, filamentous green algae and diatoms (Bunt, 1954b; Evans, 1970; Croome, 1984) (Appendix 5).

Evans (1970) used the concentration of chlorophyll *a* (the photosynthetic pigment universal in plants) as an indirect measure of algal populations in several lakes over a 15 month period. Prion Lake had a low chlorophyll *a* concentration throughout the year. In ponds with higher nutrient levels (Square Lake, Duck Lagoon), chlorophyll *a* concentration rose during spring, with increasing water temperatures and hours of daylight, to a summer high (Figure 7.7). The phytoplankton population in lakes is important to aquatic animals which feed upon them.

Freshwater fauna

The circum-subantarctic copepod *Pseudoboeckella brevicaudata* is the most abundant zooplankton species on Macquarie Island. Several other crustacean species, oligochaetes, nematodes and flatworms, as well as a variety of tardigrades, rotifers and insect larvae, have been collected from freshwater habitats (Table 7.3).

The biology and ecology of *Pseudoboeckella brevicaudata* on Macquarie Island was studied in considerable detail by Evans (1970). The species passes through six naupliar (larval) stages and five copepodid stages before reaching the adult form. On Macquarie Island, *P. brevicaudata* occurs in water bodies ranging in size from the larger lakes such as Prion Lake to small, shallow freshwater pools of less than one cubic metre volume. It sometimes appears in slightly brackish pools.

In larger water bodies such as Prion Lake the species has a distinct pattern of diurnal migration. It is uniformly distributed during the night but moves to the lake bottom during the day, a pattern presumably reflecting a similar migration in the phytoplankton on which it feeds. A similar diurnal migration occurs in the cladoceran *Daphnia carinata*, the only other planktonic crustacean on the island. In shallow ponds and lakes such as Square Lake the pattern of migration is different. *Pseudoboeckella brevicaudata* occurs throughout the lake at night but during the day lives amongst the dense mats of algae and larger aquatic plants around the lake margin. It appears that *P. brevicaudata* has a need for shelter from light. On Macquarie Island the more-deeply pigmented forms of the species occur in the most transparent water bodies. It has been suggested that such increased pigmentation may protect copepods from the greater solar radiation penetrating very clear waters.

In oligotrophic Prion Lake, Evans (1970) found that lower densities of *P. brevicaudata* were present, and the animals were smaller than in shallower ponds with higher nutrient status, such as Square Lake and Duck Lagoon. *Pseudoboeckella brevicaudata* is a non-selective filter-

Table 7.3. *Occurrence of major aquatic animals and plants in five fresh-water bodies on Macquarie Island.*

	Prion Lake	'S' pond[a]	Square Lake	Duck Lagoon	'F5' pond[a]
Copepoda					
Pseudoboeckella brevicaudata	+	+	+	+	+
Thalastrid sp.	+	+	+	+	+
Cladocera					
Daphnia carinata	+	–	+	+	+
Alona diaphana	–	+	+	+	+
Alona quadrangularis	–	+	+	+	+
Macrothrix hirsuticornis	–	+	+	+	+
Chydorus poppei-barroisi (group)	–	+	+	+	+
Acari					
One species	+	+	+	+	+
Oligochaeta					
Marionina sp.					
Several species (not identified)	+	+	+	+	+
Turbellaria					
Several species (not identified)	–	+	+	+	+
Anthophyta					
Myriophyllum triphyllum	rare	–	+	+	–
Callitriche antarctica	–	–	+	–	–
Ranunculus biternatus	–	+	+	–	–
Chlorophyta					
Rhizoclonium sp.	+	+	+	+	+
Spirogyra sp.	–	+	+	+	+
Staurastrum sp.	+	–	–	–	–

In addition, tardigrades, nematodes, rotifers and diptera larvae were present in collections.

[a]'S' pond = plateau pond illustrated in Figure 7.2; 'F5' pond = pond on coastal terrace near Duck Lagoon.

Data from Evans (1970).

feeding herbivore. In Prion Lake, availability of phytoplankton as a food source appears to limit the size of both individuals and the population.

On Macquarie Island, populations of *P. brevicaudata* are present throughout the year. There is no dormant phase and eggs can hatch at any time of the year. However, seasonal variations and variations between lakes in temperatures and food availability had important effects on the timing of the breeding cycle of *P. brevicaudata*, leading to variations in hatching time, growth rate and abundance of various stages of the life cycle.

By contrast, on Heard Island, the small ponds inhabited by the species freeze throughout the winter and a dormant phase is imposed. On neither Heard nor Macquarie Islands do the water bodies in which the species occurs ever even approach the upper lethal temperature for the species of 25–28 °C.

Whether the species has any major predators on the island is unknown. Evans was unable to find any potential or actual predators of *P. brevicaudata* in Prion Lake, although in the course of laboratory studies he observed planaria (abundant amongst mats of algae and larger aquatic plants) feeding on eggs and nauplii. There are no freshwater fish on Macquarie Island.

Pseudoboeckella brevicaudata is the only known species of subantarctic copepod which is circumpolar in range. Typically an inhabitant of cold freshwater, it is widely distributed in the subantarctic zone, occurring both on continental land masses (Punta Arenas and Tierra del Fuego, South America, the Falkland Islands and Iles Kerguelen) and on truly oceanic islands such as Macquarie and Heard Islands.

Since no stage of the life cycle can withstand immersion in seawater, its successful colonisation of the subantarctic zone and spread to subantarctic islands must have been accomplished by long-distance aerial dispersal. A juvenile giant petrel shot on Macquarie Island was found to have immature *Pseudoboeckella brevicaudata* stages among filamentous algae attached to one foot. Since the copepod is common in algal mats of ponds it is likely that eggs and other stages of the life cycle are collected along with algae. The egg is the only stage of the life cycle which can survive out of water, probably due to the presence of an outer chitinous membrane. Evans showed that eggs survive up to ten days at a relative humidity of 92%. Successful dispersal probably required algal material held close to a bird's body in flight under conditions of maintained high humidity. In an area where long-distance flights are common and the winds strong, this means of dispersal for *P. brevicaudata* eggs and resistant stages of other freshwater organisms does not appear at all fanciful.

8

The island's birds

A complete list of birds recorded from Macquarie Island is given in Appendix 10. Only those bird species which have been intensively studied on Macquarie Island and on which most has been published are discussed below.

Among the most spectacular sights on Macquarie Island are vast penguin colonies some containing hundreds of thousands of birds (cover illustration). Penguins are some of the many bird species which breed on the island but spend most of their lives at sea. Few bird species on the island are entirely terrestrial. Several reached the island only relatively recently (Tables 8.1, 8.2).

Two endemic bird species became extinct during the late nineteenth century: the Macquarie Island parakeet (*Cyanorhamphus novaezelandiae erythrotis*) and the flightless Pacific banded rail (*Rallus philippensis macquariensis*) (Hamilton, 1895). They almost certainly died out because alien predator species were introduced to the island.

When Macquarie Island was discovered, parakeets were abundant, feeding on invertebrates among stranded seaweeds. They seem to have

Table 8.1. *Self-introduced birds on Macquarie Island.*

Species	Year of introduction	Comments
Starling[a]	Unknown	Introduced to New Zealand 1862 and Australia 1863[b]
Redpoll[a]	Unknown	Introduced to New Zealand 1868[b]
Mallard[a]	First record from island 1947. May have been present earlier	Introduced to Australia and New Zealand in nineteenth century. Interbreeds with native Black duck

[a] Breeding population established on island by 1987.
[b] Dates of introduction to Australasia from Williams (1953).

Table 8.2. *Bird species deliberately introduced to Macquarie Island.*

Species	Year of introduction	Comments
Wekas	1867	Deliberately introduced as food for sealers (Cumpston, 1968: 62). Feral breeding population established
Fowls	1915	Food for meteorological station staff (Cumpston, 1968: 279)
	1955[a]	ANARE as food source
Ducks	1915	Food for meteorological station staff (Cumpston, 1968: 279, 287)
	1955[a]	ANARE for food
Geese	1955	ANARE for celebratory meals

[a] Reintroduction.

remained abundant until about 1880 when their numbers declined rapidly. The species was apparently extinct by 1891 (Hamilton, 1895), an extinction variously laid at the feet and mouths of feral cats (Mawson, 1943), feral cats and dogs (Taylor, 1955a) and wekas (Law & Burstall, 1956). For nearly 70 years a combination of humans, feral cats and sealers' dogs appears to have had no dramatic effect on the parakeet population. In the late 1870s, wekas and rabbits were released on the island. Cats, rabbits, and possibly wekas, increased in numbers and parakeets became extinct during the 1880s. Taylor (1979) thought that a series of events was involved in the extinction of parakeets. Introduction of rabbits, whose rapidly growing population provided additional winter food for cats and wekas and hence allowed increases in populations of these predators, led to intolerable predation pressure on parakeets and caused their eventual extinction.

Flightless banded rails seem to have become extinct at about the same time. They were abundant in 1879 but extinct, or very much reduced in numbers, by 1894 (Hamilton, 1895). Predation by wekas and black rats was probably responsible for the rail's decline (Vestjens, 1963). Three museum specimens of partial rail skulls exist.

Penguins
Penguins account for over 90% of total bird biomass in Antarctica and the subantarctic (Croxall & Prince, 1980) and are by far the most numerous birds on Macquarie Island. Four species breed there: royal

penguins (*Eudyptes schlegeli*), king penguins (*Aptenodytes patagonicus*), gentoo penguins (*Pygoscelis papua papua*) and rockhopper penguins (*Eudyptes chrysocome*).

Royal penguins

Royal penguins apparently breed only on Macquarie Island but regularly visit Heard Island. Although royal penguins (Figure 8.1) are regarded as a species separate from macaroni penguins (*Eudyptes chrysolophus*), which breed on many subantarctic islands, criteria for separation of the species are not absolutely clear. Measurements of characters such as bill length, width and depth, flipper length and width, and body weight have been used to differentiate both species and sexes. Although macaroni penguins usually have black faces, and royal penguins white, intermediate facial colours occur in royal penguins on Macquarie Island. White-faced birds are common in colonies on the east coast and black-faced birds common in colonies on the west coast (Shaughnessy, 1975). The darkest-faced royal penguins on Macquarie Island are strikingly similar to macaroni penguins. There is no marked sexual dimorphism in royal penguins, although males may be slightly heavier than females and have larger, deeper bills (Woehler, 1984).

Fifty-six royal penguin colonies were located on the island during a survey made in 1984 (Copson & Rounsevell, 1987). Colony size ranged from 75 breeding pairs to over 160 000 breeding pairs. Royal penguin colonies mainly occur on the coast, although some are well inland, such as one over 200 m asl on the slopes of Mount Elder. On the west coast many royal penguin colonies occur on very rocky uneven ground. An enormous colony at Hurd Point covers about 67 000 square metres. Royal penguins on the Bishop and Clerk Islands number 'in excess of 1000 birds' (Lugg, Johnstone & Griffin, 1978).

If the ground is even, nest density of royal penguins is relatively uniform, providing a basis for estimating populations. A nest density of 2.4 nests per square metre, derived from study of a colony at Sandy Bay, was used by Copson & Rounsevell (1987) to estimate populations in some colonies. Other colonies were accurately surveyed. The breeding population of royal penguins on Macquarie Island is somewhere between 810 000 and 960 000 breeding pairs – one of the greatest concentrations of seabirds in the world.

Royal penguins on Macquarie Island were exploited for oil during the nineteenth century. Birds at the Nuggets were the most heavily exploited because their colonies were closest to the works already established for

Figure 8.1. Royal penguins making their way along Finch Creek.

boiling down elephant seals (Figure 2.1). Blake (Mawson, 1943) surveyed the Nuggets colonies in 1913 and found that royal penguins occupied 35 000 square metres. In 1984 the Nuggets colonies covered 40 000 square metres (Copson & Rounsevell, 1987). Cessation of boiling down penguins for oil could account for the 16.5% increase in area of the colonies between 1913 and 1984. The huge colony at Hurd Point, estimated by Blake in December 1912 (Mawson, 1943) as covering over 69 000 square metres, increased in area by only 1.1% between 1912 and 1984.

Royal penguins are absent from Macquarie Island during winter. The breeding cycle is highly synchronised (Carrick, 1972). The first breeding birds return to the island in mid-September and lay eggs in October. All surviving chicks are fledged by mid-February. Adult birds then leave the island for a short period and return to moult in March and April. All royal penguins leave the island by early May to spend winter at sea (Warham, 1971).

The breeding population of royal penguins has remained fairly constant since detailed studies began. Only about 10% of birds survive to 6 years. Females are 5–11 years old when they first lay eggs. Annual survival of breeding birds is 80–90% (Carrick & Ingham, 1967).

Royal penguins which breed successfully are earliest to return to the island, older, and weigh more than other royals when they come ashore. Breeding success appears to be determined by experience and social status. These factors determine which birds are best fed, and the best-fed birds come ashore earliest. Feeding success and social status together determine access to better nesting sites and mate selection (Carrick & Ingham, 1967; 1970). Feeding success is important as females stay ashore without feeding during egg-laying and the first period of incubation. Males arriving on the island to breed weigh up to 7 kg. Females weigh 4.2–6.3 kg. Females lighter than 4.8 kg rarely breed successfully.

Gentoo penguins

Gentoo penguins are widespread in the subantarctic, breeding on Macquarie Island, the Falkland Islands, Iles Crozet, South Georgia, the South Shetland, South Orkney and South Sandwich Islands, and on the Antarctic peninsula as far south as 65°. Only 2% of the total gentoo penguin population occurs on Macquarie Island (Robertson, 1986).

Morphological variation within gentoo penguin populations is considerable and has led to description of several subspecies. There has been confusion as to how many subspecies are represented over the birds' wide geographical range. Stonehouse (1967, 1970) recognised two subspecies: a northern population (*Pygoscelis papua papua*), including Macquarie Island birds, and a southern population (*Pygoscelis papua ellsworthi*) in the South Orkney Islands. Stonehouse's studies do not support earlier recognition of another subspecies, *Pygoscelis papua taeniata*, on Macquarie Island.

On Macquarie Island, gentoo penguins nest on flat or gently sloping coastal sites up to about 70 m asl among *Poa foliosa* tussocks (Figure 8.2) and occasionally in open areas, avoiding areas of possible elephant seal disturbance. They build widely spaced grassy nests, and colonies are usually well-separated from those of other penguin species. Nest-building has considerable temporary impact on the surrounding vegetation (Chapter 6).

Figure 8.2. Gentoo penguin colony in *Poa foliosa* tussocks, Hasselborough Bay.

Most gentoo penguin colonies are on the island's north-west coast. Colonies on the west coast are larger than those on the east coast (Robertson, 1986). Allowing for a failure rate of 5% in hatching (Croxall & Prince, 1979), Robertson (1986) calculated that approximately 5000 pairs formed the breeding population on Macquarie Island during the 1984–85 summer. Breeding success varies between colonies (Reilly & Kerle, 1981; Robertson, 1986), usually being higher in larger colonies.

Gentoo penguins feed mostly on fish caught relatively close inshore (Croxall & Prince, 1980) and can dive to more than 100 m (Conroy & Twelves, 1972). Macquarie Island's west coast offers feeding grounds some 60% greater than feeding grounds off the east coast due to submarine topography.

Rockhopper penguins

Rockhopper penguins (Figure 8.3) breed in subantarctic and more northern localities (Falkland Islands, Tierra del Fuego, Tristan da Cunha, Gough Island, Marion Island and Iles Crozet, Iles Kerguelen and Iles Amsterdam, Ile St Paul, Heard Island, Macquarie Island, and the Auckland and Campbell Islands).

On Macquarie Island, most large rockhopper penguin colonies are on

Figure 8.3. Rockhopper penguins.

the west coast. Colonies may adjoin, and even intermingle with, royal penguin colonies. Rockhopper penguins generally nest on talus slopes or in mixed colonies, making population estimates difficult. In 1963 the population was estimated at several hundred thousand and more recently at 500 000 pairs (Warham, 1963; Copson, unpublished data).

Gwynn (1953a) studied egg-laying and incubation in rockhopper, macaroni (including royal) and gentoo penguins on Heard and Macquarie Islands. Rockhopper penguins on Macquarie Island lay about two weeks earlier than those on Heard Island. Of the two eggs laid, the first, smaller, egg is usually wasted.

Male rockhopper penguins arrive on Macquarie Island about 15 October each year, returning to the same nest site where they moulted some five-and-a-half months earlier. Females return between 24 October and 3 November. Laying occurs between 8 and 18 November. Incubation starts only after the second egg is laid (hence the waste of the first egg) and takes 33–34 days (Gwynn, 1953a; Warham, 1963).

During winter 1985, rockhopper penguin diet was studied in two small colonies at Garden Cove. Analysis of gut contents shows that cephalopods, euphausiids, fish and amphipods are eaten. Euphausiids occur in 99% of gut samples, fish in 70%, amphipods in 26% and cephalopods in 15%. There appears to be little change in relative percentages of food items from season to season.

Variation in rockhopper penguin populations on Macquarie Island have not been studied due to difficulties in determining numbers. Moors (1986) described a decline of up to 50% in numbers on Campbell Island since the first counts were made there in the 1940s. Reasons for the decline in the Campbell Island population are not known. Increased predation by introduced animals and disturbance by humans or sheep have all been discounted. Predation by brown rats (*Rattus norvegicus*) has been ruled out (Westerkov, 1960). Feral cats on Campbell Island feed mainly on rats. Studies of rockhopper penguin biology continue on Campbell Island and similar studies are desirable on Macquarie Island.

King penguins
Two colonies of king penguins existed on Macquarie Island in 1810. The birds were exploited for their oil and the colony on the isthmus was obliterated by 1840 (Mawson, 1943). The largest population, fortunately, was at Lusitania Bay. This colony, although probably severely depleted, was still large, possibly until about 1895 (Falla, 1937), but by 1911 had been reduced to about 5000 birds (Mawson, 1943). The Lusitania Bay colony (cover illustration) has been unmolested since exploitation of penguins ceased and by 1956 the colony had increased to more than 10 000 pairs (Law & Burstall, 1956). Since counts were first made in 1930, there has been a dramatic increase in numbers in the Lusitania Bay colony (Rounsevell & Copson, 1982). By 1980, there had been a 78-fold increase in annual chick production, the most dramatic increases occurring since 1960. By 1975, all available flat land near the colony was occupied by breeding birds. The colony has gradually extended into tall tussock grassland and onto landslip sites. Rounsevell (1983) reported that the number of chicks in the Lusitania Bay colony was doubling every four years, an exponential rate of increase. Rounsevell pointed out that the Lusitania Bay population has been growing at a steady rate from 1930 onwards. Food supplies have not determined population growth.

Figure 8.4. King penguins and chicks in a colony at Sandy Bay.

Population growth has been determined by the maximum reproductive rate of the species (Rounsevell, 1983).

In 1975, king penguins established a new colony at Sandy Bay, 17 km north of the Lusitania Bay colony. The population in this new colony has been constantly monitored and all chicks have been flipper banded.

A king penguin pair successfully raises a maximum of two chicks (Figure 8.4) every three years (Stonehouse, 1960). A 1980 count of 47 000 chicks in the Lusitania Bay colony implies a minimum 70 000 breeding pairs, although egg predation, flooding of the colony and other factors led Copson to suggest that the colony may contain as many as 100 000 breeding pairs (Copson, unpublished data). The total population in the colony in 1980 was estimated at between 218 000 and 250 000 birds (Rounsevell & Copson, 1982).

King penguin diet has been studied using a water-offloading technique which makes the birds regurgitate all food. Samples were collected at monthly intervals during 1985 from ten birds. The average amount of food collected from the birds, 923 g, represents a mean 8.3% of their body-weight. Similar dietary studies have been conducted on both royal and rockhopper penguins.

King penguins appear to feed seldom if at all during egg-laying and incubation. As chicks hatch and grow, adults bring increasing amounts of food ashore, reaching a peak in April. In winter, chicks maintain themselves with little growth until the following spring and less food is brought ashore. The increase in energy demand by chicks the following spring is reflected in a sudden increase in the amount of food brought ashore by adults in September.

Fish and cephalopods are very common in the diet. Fish occur in 98% of samples and account for more than 90% of food caught. Although cephalopod remains occur in many samples, they appear only in small amounts and form less than 3% of the king penguin diet (Hindell, 1988).

Giant petrels

Northern giant petrels (*Macronectes halli*) breed only on islands between the Antarctic and Sub-tropical convergences. Southern giant petrels (*M. giganteus*) (Figure 8.5) breed only on islands near the Antarctic Convergence and southwards towards Antarctica. Just north of the Antarctic convergence, in the zone of overlap which includes Macquarie Island, the species breed sympatrically, remaining separate because of differences in nesting habits and breeding times.

On Macquarie Island, southern giant petrels nest in colonies in exposed sites. Northern giant petrels nest singly or in loose aggregations in sheltered sites (Carrick & Ingham, 1970). Large congregations of both species remain more or less in the same areas from year to year, although possibly shifting nesting sites.

On Macquarie Island, specific isolation of the two giant petrels is maintained by differences in egg-laying periods. *Macronectes halli* lays from about 11 August to 6 September; *M. giganteus* from 27 September to 19 October. Attempted interbreeding between species has been observed but there is no evidence that any such attempt has been successful (Johnstone, 1978).

There are about 1000 pairs of *Macronectes halli* and 4000 pairs of *M. giganteus* on the island (Johnstone, 1977). About 10% of *M. giganteus* are 'white phase' birds, the adults white with a few scattered dark feathers. The rest are 'dark phase': dark grey as fledglings, becoming paler and brownish with age (Shaughnessy, 1971).

Figure 8.5. Nesting southern giant petrels near Boiler Rocks.

Thousands of chicks of both species have been banded on the island. Specimens of *Macronectes halli* banded on Macquarie Island have been recovered in South Africa, South Georgia, Chile, Argentina, New Zealand, southern Australia and Fiji. Some birds banded on Heard Island have turned up 14 500 km from their colonies less than two months after leaving (Howard, 1954).

Unguarded giant petrel eggs are often lost to cats and skuas.

Giant petrels are scavengers and predators on land and at sea. The diet of both species is similar (Johnstone, 1977) although *M. giganteus* appears to eat slightly more shoreline-caught food. Both species scavenge dead seals and placental remains and kill both adult penguins and chicks. Favoured hunting places allow the big birds a long straight path for take-off. Their effect on penguin populations is probably slight, although their hunting methods are spectacular.

At sea they catch a variety of food. They feed on floating carrion, ship's refuse and catch small birds in flight. There are no pronounced differences in diet between birds from different subantarctic islands or Antarctica (Mougin, 1968; Johnstone, 1977; Voisin, 1968; Downes, Ealey, Gwynn & Young, 1959; Conroy, 1972).

Table 8.3. *Burrow-nesting birds breeding on Macquarie Island.*

White-headed petrels – Absent in winter. Breed November–February, in tall tussock grassland 100–300 m asl. Occasional mixed colonies with sooty shear-waters. About 8000 burrows on island. Heavy predation by cats, rats, wekas. Do not tolerate rabbits.

Blue petrels – Nest on rock stacks in rock crevices under tussocks or in *Cotula plumosa* and *Colobanthus muscoides*. Present all year. Eggs October. No chicks ever seen. Population 500–600.

Antarctic prions – Nest 100–420 m asl in short grassland and herbfield on sheltered slopes or in low-altitude areas dominated by *Acaena magellanica*. May burrow in *Azorella macquariensis* in feldmark. About 49000 breeding pairs. Eaten by cats and skuas. Cats kill 26000 per year. Often nest in existing rabbit burrows.

Fairy prions – First record 1979. About 40 breeding pairs. May compete with blue petrels for burrow sites.

Sooty shearwaters – Colonies along coast in tussock grassland. Also in short grassland and herbfield; 100–150 m asl. May compete with rabbits for burrows. Eaten by skua, cats, wekas.

Common diving petrels – First record 1979. Breeding population about 20 pairs. Blue petrels occupy diving petrel burrows and species may compete for burrow sites.

For a detailed treatment of burrow-nesting birds breeding on Macquarie Island refer to Brothers (1984) and Warham (1967).

Burrow-nesting petrels

Macquarie Island's original bird fauna included large numbers of penguins, surface-nesting land and seabirds and burrow-nesting petrels. Cats, wekas, black rats and rabbits have affected the island's bird populations, particularly those species which nest in burrows. Tables 8.3 and 8.4 give details of burrow-nesting birds on Macquarie and their present status. It appears that only one species of burrow-nesting petrel (the grey petrel) has ceased breeding on the island.

White-headed petrels

White-headed petrels return to their nests in November. Chicks hatch early in February and all birds leave the island in May.

Burrow-nests of white-headed petrels have been recorded at altitudes between 100 m and 300 m (mostly between 150 m and 250 m) in tall tussock

Table 8.4. *Burrow-nesting birds visiting Macquarie Island but not breed-ing.*

Thin-billed prions	Fulmar prions
Soft-plumaged petrels	Grey petrels[a]
Short-tailed shearwaters	Grey-backed storm petrels

[a] Early records indicate that grey petrels once bred on island.

grassland or short grassland communities, occasionally in mixed colonies with sooty shearwaters. All nests are on sheltered slopes.

Variations in burrow density make estimates of the population on the island difficult. Brothers (1984) estimated that there are about 8000 burrows.

White-headed petrels are subject to heavy predation. In tall tussock grassland, cats, rats and wekas are the main predators. In short grassland and feldmark, cats appear to be major predators.

Blue petrels

Brothers (1984) located six colonies of blue petrels on the island, all on rock stacks. Most burrows are 10–14 m asl, amongst *Cotula plumosa* and *Colobanthus muscoides* with occasional *Poa foliosa* tussocks.

Blue petrels have been found nesting in rock crevices, under tussocks of *Poa foliosa*, and occasionally on the surface, throughout almost all the year.

Brothers studied one readily accessible colony at Green Gorge in an attempt to establish the breeding cycle. Although eggs have been found, no chicks have been observed. There are about 500–600 blue petrel pairs on the island. The apparent total breeding failure of the species presents a problem as to why any blue petrels remain on the island. Predation by skuas may be the major cause of breeding failure. Remains of blue petrels are regularly found in skua territories (Jones, 1980; Brothers, 1984; Merilees, 1971).

Antarctic prions

Antarctic prions return to the island in mid to late spring. They nest in short grassland and upland herbfield on sheltered slopes and in gully banks. Some burrows occur on coastal slopes. At lower altitudes, prions favour areas dominated by *Acaena* species. In feldmark, burrows tend to be between clumps of *Azorella macquariensis* rather than in or under cushions of the plant as on other islands (Falla, 1937; Tickell, 1962).

17% to 44% of prion burrow-nests are associated with rabbit burrows, the birds exploiting existing rabbit burrows from which they either dig offshoots within about 50 cm of the burrow entrance or nest deep within the rabbit burrow system.

There are about 49000 breeding pairs of Antarctic prions on the island. Their nest distribution parallels that of white-headed petrels. Brothers (1984) reported a hatching success of 81.8%, with all chicks fledging. Predation is the major cause of prion mortality. The abundance of rabbits in the areas favoured by prions may alleviate predation pressure on the birds. Nevertheless, analysis of cat gut and scat contents suggests that cats may kill up to 26000 prions annually, although the true figure is likely to be lower (Brothers, 1984).

Sooty shearwaters

Sooty shearwater breeding colonies are scattered along the coastline amongst stands of *Poa foliosa* on slopes or in short grassland or herbfield. Most colonies are found between 100 and 150 m asl. Tussocks of *Poa foliosa* in the colonies often appear darker green than tussocks outside the colony limits, probably due to manuring by the birds with addition of nitrogen (Brothers, 1984).

Adult shearwaters are eaten by skuas and cats. Wekas, rats and rabbits interfere with burrows.

Southern skuas

Southern skuas nest on the coastal terrace and plateau of the island (Jones and Skira, 1979). Skua diet is diverse and varies from place to place in the subantarctic. On both South Georgia (Stonehouse, 1956) and Campbell Island (Bailey & Sorensen, 1962), they eat brown rats; on Iles Kerguelen, rabbits (Lesel & Derenne, 1975). On Signy Island, they use every food source except terrestrial and freshwater invertebrates (Burton, 1968). Southern skuas in the subantarctic are both scavengers and birds of prey. They feed on carrion, live animals such as rabbits and other forms of food such as milk oozing from the mammary glands of lactating southern elephant seals, rabbit dung and human refuse.

On Macquarie Island, skua diet has been studied by analysis of ejecta (Jones & Skira, 1979). Rabbits and burrow-nesting petrels are eaten by skuas nesting on the plateau. Penguin feathers and southern elephant seal hairs also occur in ejecta. In feldmark areas, we have watched skuas turning over polsters of the moss *Ditrichum strictum*, perhaps foraging for small invertebrates such as worms and slugs.

Ejecta from skuas with territories on the coastal terrace contain large amounts of penguin eggshell, remains of penguins and elephant seal hair. Skuas kill rabbits weighing up to 1.2 kg or more. Of rabbits eaten, 82% weigh under 0.45 kg and 98% under 0.75 kg (Jones & Skira, 1979). We have seen a fully-grown rabbit carried aloft by a skua, only to be dropped. Johnston (1973) reported successful skua kills of rabbits up to 1.3 kg and unsuccessful attempts at catching rabbits up to 2.8 kg. Successful introduction of myxoma virus into the rabbit population means that larger rabbits are now being caught by skuas due to sickness of rabbits.

Skuas and giant petrels were once the only effective predators of terrestrial prey. Changed predator/prey relationships involving both skuas and giant petrels have occurred since the introduction of rabbits, cats, rats and wekas.

The first skua eggs are laid on Macquarie Island in early October and the first chicks hatch in mid-November. Most eggs hatch by mid-December and by mid-January the earliest-hatched chicks have begun to fly. The breeding season coincides with the presence of maximum numbers of young rabbits and the breeding cycles of penguins and burrow-nesting petrels.

Small piles of skua gizzard stones are commonly found on the plateau. Skuas were responsible for distributing small pieces of pumice over the plateau following arrival on the island's coast in 1963 of a raft of pumice. Simpson (1965) showed that while pumice was common on island beaches many skua gizzard stone regurgitations consisted almost entirely of pumice and skua nests contained pumice, as did nests of royal penguins.

Albatrosses

Wandering albatrosses

Wandering albatrosses are often considered to have been numerous on the island, basically on anecdotal evidence that they were killed by sealers for food (Cumpston, 1968). Numbers on Macquarie Island have probably always been small compared with large populations of the species on other subantarctic islands (e.g. an estimated 2000 on Marion Island). Macquarie Island has probably always been a marginal habitat for them. There appears to have been a decline in the island's breeding population over recent years and it is now very small. Croxall (1979) described a decrease in numbers of wandering albatrosses on South Georgia between the late 1960s and the late 1970s.

Some wandering albatrosses move from their natal island to breed on other islands, making it difficult to assess overall populations on any island. A banded bird's failure to return to Macquarie Island for two or more years, or failure of observers to note it even if present, leads to its classification as 'disappeared'. 'Disappearance' has usually been taken as implying death (Tickell, 1968) or breeding on another island. A male, banded on Macquarie Island in 1967 as a non-breeding adult, bred on Heard Island in 1980 (Johnstone, 1980). Mougin (1977) reported a bird hatched on Ile de la Possession breeding on Marion Island.

Fledging successes on Macquarie Island of 45% in 1965–68 (Carrick & Ingham, 1970) and 53% in 1974–78 (Tomkins, 1985a) are not greatly different from those recorded on Bird Island near South Georgia (59%) (Tickell, 1968; Croxall, 1979) and on Ile de la Possession (64%) (Fressanges du Bost & Segonzac, 1976).

Death rates of adult wandering albatrosses of 5–10% on Bird Island (South Georgia) (Tickell, 1968; Croxall, 1982) and 3.6% on Ile de la Possession (Barrat, Barre & Mougin, 1976) are much lower than the 21.3% reported for Macquarie Island (Tomkins, 1985a), but some Macquarie Island birds posted as 'missing', presumed dead, have subsequently returned to the island. If the eight pairs which 'disappeared' from the island in 1974–78 bred on other islands, the mortality rate for the Macquarie Island population is only 9.8%, a rate similar to those recorded elsewhere. Carrick & Ingham (1970) estimated annual mortality of breeding birds on Macquarie Island in the period 1965–68 at 5.4%.

While the breeding population on Macquarie Island is small, the colony is visited, at least during the early part of the breeding season, by a number of non-breeding birds. Young males and females visit occasionally and such visits probably become more frequent and regular as the birds become older. Older non-breeding males visit more frequently than older non-breeding females. Older females visit more frequently than young females. This pattern of behaviour is probably related to establishment of territorial claims (as shown by display nest-building) and eventual mate selection (Richdale, 1950; Tickell, 1968).

No one knows where wandering albatrosses from Macquarie Island feed, but they travel long distances. Two banded birds were recaptured during a single breeding season, one near Tasmania (1600 km from the island), the other off the coast of New South Wales (2240 km from the island). At least some of the birds move to the east coast of the Australian mainland to feed (Tomkins, 1985b). During the breeding season, long absences from the island occur around the time of full moon. These are probably related to difficulties in catching cephalopods at such times, the cephalopods and their prey not migrating as high in the water column during lighter periods (Wickstead, 1976). Non-breeding males and females leave the island overnight.

The large population (about 2000 birds) of wandering albatrosses on Marion Island formed the basis of an interesting study of the species' influence on vegetation in the vicinity of nest sites (Croome, 1972). An expedition to Marion Island in 1965–66 counted about 3000 wandering albatross nests, not all occupied and not all of the same age. Almost all nests occurred in coastal mire sites below 100 m asl. Croome made detailed studies of vegetation in the vicinity of a number of nests and found that drainage from nests contained considerable amounts of available nitrogen resulting in pronounced stimulation of growth of *Poa cookii* downslope from nests, the nitrogen being carried by slowly moving ground water. Croome concluded that wandering albatross guano is the greatest source of nitrogen added to Marion Island's terrestrial ecosystem by birds. Even nests which had apparently not been used for five to six years showed much higher nitrogen concentrations beneath them than in surrounding areas. No such studies have been made on Macquarie Island.

Light-mantled albatrosses
Light-mantled albatrosses (Figure 8.6) have been studied on Macquarie Island since 1970. Their nests are scattered along the coast on steep slopes and cliffs. The major area for study of the species has been at the far northern end of the island, with a control colony at Bauer Bay being monitored for studying possible effects of human disturbance. The breeding population is estimated at 500–700 pairs (Kerry & Colback, 1972).

Once a pair bond has been established, light-mantled albatrosses tend to remain together for life. One pair breeding in 1954–55 was still together in 1975–76. Pairs return to the same locality each year, and often re-use old nests. There is evidence that chicks return to their hatching sites to breed when they eventually reach breeding status.

Light-mantled albatrosses leave the island in winter. Members of established breeding pairs arrive on the island at about the same time early in October. Courtship extends over about 10 days and, after mating, the female returns to sea, leaving her mate to construct a nest at the chosen site. She returns several days later to lay a single egg and then departs, leaving the male to carry out the first stages of incubation. Incubation is later shared between partners. Eggs are laid in late October to early November and first chipping of eggs occurs in late December to early January. Chicks take 3–5 days to emerge from the egg and fledged chicks

Figure 8.6. Light-mantled albatross on nest, Sawyer Creek.

leave the nests about mid-May to mid-June. The period from hatching to final departure of chicks from the island is about 144–153 days.

If the single egg laid is damaged, it is not replaced. Breeding success, measured by percentage of chicks fledged, was 52% during the breeding seasons of 1970–71 and 1980–81. If a chick is successfully reared, adults do not breed the following season. The same phenomenon occurs in wandering albatrosses. On Macquarie Island, only 33% of successfully breeding light-mantled albatrosses return to breed two years later and a further 18% in the third year following successful breeding. Pairs are thus capable of producing at most one chick every two years but the average is one chick every three or four years.

There seems to have been a steady decline in breeding success of light-mantled albatrosses on Macquarie Island since 1971–72. The trend has been the same in three colonies at the northern end of the island and the colony at Bauer Bay monitored for human disturbance. Marking eggs and checking and banding fledglings may adversely affect the birds at critical stages of the breeding cycle.

Only preliminary dietary studies have been carried out on Macquarie Island, showing that cephalopods form a large part of the diet.

Cats disturb nesting birds and young chicks. Cats, or cat tracks, are common around and near accessible nesting sites during the breeding season. Light-mantled albatrosses will not retrieve an egg or young chick displaced from the nest, so disturbance must be minimal for breeding to be successful.

Recoveries of banded birds show that light-mantled albatrosses are potentially long-lived. Of 66 birds banded between 1954 and 1964, 11 have been observed 15 or more years later. The oldest banded birds were then 27 years old. Until 1984 only two birds banded on Macquarie Island as chicks had joined the breeding population there, a female of eight years and a bird of unknown sex at six years.

The movements of birds once they leave the island are almost unknown. Two banded fledglings were found, four months after leaving the island, on the North Island of New Zealand. Young birds return to Macquarie Island after about six years (K. Kerry, unpublished).

Other albatross species

Both black-browed albatrosses (*Diomedea melanophrys*) and grey-headed albatrosses (*D. chrysostoma*) have small breeding populations on the island. There are three known colonies of black-browed albatrosses. The one on North Head seems headed for extinction. No chicks have fledged from it since the early 1970s. At the same time the other colony on the main island, in the south-west corner, has been increasing in size. The third colony is on the Bishop and Clerk Islands and appears to be the largest. Mackenzie (1968) reported 25+ pairs and Lugg, Johnstone & Griffin (1978) give 44+ pairs for the colony. With a total population of 20–30 pairs on Macquarie Island, that population is marginal. Grey-headed albatrosses are more common, numbering 80–100 pairs in a single colony in the south-west corner of the island.

The two species differ in nesting habits. Black-browed albatrosses nest in tall tussock grassland; grey-headed albatrosses prefer more rocky exposed areas (Copson, 1988).

Kelp or Dominican gulls

Kelp or Dominican gulls (*Larus dominicanus*) have a circumpolar distribution in the southern hemisphere. They utilise many food resources in different ways under different environmental conditions. They have virtually no competitors for food, their diet being chiefly intertidal marine molluscs. Tidal conditions and wind are important in limiting where and when they can feed. During low tides in calm weather the number of gulls feeding inshore is proportional to the lowness of the tide, allowing foraging on chitons, limpets and other molluscs. During high tides and high seas they feed on adult and larval kelp flies among rotting beach-stranded kelp, flies being taken directly from the kelp or from the beach sand or water (Merilees, 1984a).

Table 8.5. *Summary of food items in regurgitation of Dominican gulls.*

Food items	Winter	Spring
Marine items (e.g. molluscs, squid, isopods, fish)	42.5%	58.7%
Terrestrial animals (e.g. scavenged dead seals, kelp flies)	42.4%	26%
Plant material	3.7%	5.7%
Sand, pebbles	11.4%	9.6%

Data from Merilees (1984a).

As scavengers, dominican gulls compete directly with giant petrels and skuas and are subordinate to both. Gulls often congregate in inland localities, particularly in wet mire sites, possibly feeding on slugs, moths and moth larvae and other invertebrates. A summary of food items in regurgitated pellets from dominican gulls is given in Table 8.5.

Blue-eyed cormorants

Blue-eyed cormorants occur widely throughout the subantarctic. There is some confusion regarding their taxonomy. We accept *Phalacrocorax albiventer purpurascens* as the designation for the Macquarie Island population (Brothers, 1985).

Twenty-three cormorant breeding sites are known around the island. Not all are used each year. The two largest colonies, among boulders at Handspike Point and at Green Point, contain more than 50% of the total island population, estimated at 660 breeding pairs. Colony sizes range from three to 320 pairs. There are about 100 pairs on Bishop and Clerk Islands.

Cormorant nests (Figure 8.7) are largely made of guano, grass (chiefly *Poa foliosa*), *Cotula plumosa* and mud. Nests become heavily cemented with guano and are repaired as growing chicks damage them.

Predation by skuas and starvation are major causes of nestling deaths. Parents guard or brood chicks until they are well advanced but skuas are able to dislodge larger chicks, only partially covered by adults. Feral cats are regularly seen at the edge of the colony at Handspike Point and cats possibly eat chicks.

Cormorants' diet consists almost entirely of benthic fish, mostly *Notothenia magellanica* and *Harpagifer bispinis*. The juvenile pelagic stage of *Notothenia magellanica* is not caught for food. The birds dive to considerable depths. Conroy & Twelves (1972) reported dives of 25 m in

Figure 8.7. 'At the cormorant rookery'. Photograph F. Hurley, 1911. Reproduced courtesy Mitchell Library, State Library of New South Wales.

the South Orkney Islands. A feeding area of 90 km², of which 60 km² is on the west coast, would be available around Macquarie Island, assuming a diving depth of 50 m. Feeding areas would be substantially increased if areas around Judge and Clerk Islands and Bishop and Clerk Islands were utilised. Blue-eyed cormorants are not strong fliers and cannot make headway in winds greater than 20 m s⁻¹ (40 knots). Access to feeding grounds around outlying islands of Macquarie Island may be limited by frequent strong winds. Birds feeding on the east coast may have difficulty returning to their nests during storms, and rough weather may make feeding impossible.

Gentoo penguins, whose diet includes a large proportion of fish, probably compete with cormorants for food, as they are capable of diving to depths of at least 100 m, and thus have a much greater feeding area. On the basis of mean weight of birds and feeding requirements, Brothers (1985) estimated that the Macquarie Island cormorants harvest about 260 tonnes of fish annually.

Ducks

Ducks on the island have been described by Norman (1987) and Norman & Brown (1987). Apart from a single report of an unfortunate vagrant white swan which decorated a table (Rourke, 1903, in Cumpston, 1968), other records relate to Pacific black duck (*Anas superciliosa*), grey teal (*Anas gibberifrons*) and mallard (*Anas platyrhynchos*), and hybrids between native black duck and mallard. Black duck and grey teal were noticed by early visitors to the island. Mallard were not formally recorded until 1949 despite the fact that they may have been present as domestic stock as early as 1915. Field identification of various ducks (particularly hybrids) is difficult and lack of earlier reports of mallard may not necessarily mean that they were absent from the island.

The grey teal is a highly nomadic species with populations in both Australia and New Zealand. Particularly during droughts in Australia, grey teal disperse widely (Mills, 1976; Frith, 1982).

Pacific black duck have almost certainly been on the island for a long time. Raine (1822; 1824, in Cumpston, 1968), in the earliest published account of the island, distinguished between 'teal' and 'wild duck'. Nearly all reports of Pacific black duck on Macquarie Island have been from mire areas, herbfield or tussock-dominated areas.

Dietary observations on 'ducks' amount to reports of 'grazing' on short herbage, feeding along the shoreline on kelp fly larvae (Falla, 1937) or 'dabbling' in Duck Lagoon at a time when water samples collected from the feeding area contained mites, collembola, copepods, ostracods, clado-cerans and large quantities of filamentous green algae (Norman, 1987).

Self-introduced birds

The times of arrival on the island of known non-native bird species are shown in Tables 8.1 and 8.2. Some were deliberately introduced, others reached the island by natural means. Birds have excellent chances of covering the long distances from neighbouring land masses to Macquarie Island, especially if storm-driven. Not all bird species reaching the island have been able to establish breeding populations there (Appendix 10).

Redpoll, common starling and mallard, all originally northern hemisphere species, have established breeding populations on Macquarie Island after initial introduction to Australia and New Zealand in the mid to late nineteenth century when acclimatisation societies were active. Redpolls (introduced to New Zealand in 1868) form small flocks during winter and autumn and feed on seed heads of *Cotula plumosa* and *Pleurophyllum hookeri* (Warham, 1969). Starlings, introduced to New Zealand in 1862 and Australia in 1863, are very adaptable and breed in crevices and caves around the island. Flocks of up to 250 have been seen. There are no records of first arrival on the island of either starlings or redpolls.

9

Mammals – indigenous and introduced

Indigenous mammals

The only mammals indigenous to Macquarie Island are seals. The island is a land base for a large population of southern elephant seals (*Mirounga leonina*) and, at the time of discovery, supported large numbers of fur seals. The type (or types) of fur seal originally present on the island is not known, but the island is now being recolonised by subantarctic, Antarctic and New Zealand fur seals, all members of the genus *Arctocephalus*. Other seal species reach the island from time to time but do not breed there. Hooker's sea lions from New Zealand and leopard seals are regular visitors. Even solitary wandering crab-eater and Weddell seals from the Antarctic pack ice very occasionally haul out on the island.

The surrounding sea contains remnant populations of whales. Rare sightings of whales such as southern right whales and sperm whales are made from the island. Strandings of whales are rare. Cumpston (1968: 97) mentions a small whale beached on the island and rapidly reduced to a skeleton by scavenging birds. There are regular sightings of killer whales (*Orcinus orca*). This is probably the only whale species which has any detectable effect on the island ecosystem, preying on southern elephant seals and penguins.

Southern elephant seals

Southern elephant seals spend most of their lives at sea in Antarctic and subantarctic waters, visiting land to breed, moult and rest (Figure 9.1). The species occurs over a wide latitudinal range on both sides of the Antarctic convergence with breeding colonies between 40°S and 62°S. Non-breeding herds and occasional wandering individuals extend the species' total known range from the tropics (16°S) to the Antarctic mainland (78°S).

Figure 9.1. Small group of southern elephant seals ashore on Macquarie Island, Green Gorge. Dominant male (upper right), cows and pups.

Some breeding colonies of southern elephant seals (*Mirounga leonina*), such as the one on King Island in Bass Strait, were wiped out during the early 1800s and have never re-established. Extinction was also the fate of the southern elephant seal population at the species' type locality – Juan Fernandez, off Chile. In Tasmania, archaeological studies of midden deposits along the north-western coast show that aboriginal Tasmanians regularly hunted southern elephant seals and possibly drove at least one breeding colony into extinction some 1500 years ago (Jones, 1987). Today only occasional wandering southern elephant seals reach Tasmania. Two pups are known to have been born on Tasmanian coasts this century; one in 1958, one in 1975 (Bryden, 1983).

Almost half the world's population of southern elephant seals, estimates of which vary widely, is based in waters around South Georgia. They were commercially exploited until 1964. The Iles Kerguelen–Heard Island population accounts for about 30% of total world numbers. Smaller, but still substantial, populations are based around Macquarie Island, Chatham Islands, and the Prince Edward Islands. Scheffer (1958) details all then-known breeding and non-breeding herds of *Mirounga leonina*.

Laws (1953) wrote:

> Until more detailed comparative work has been done there is no justification for the recognition of more than one species of elephant seals in the Southern Hemisphere, although several have been proposed. A recent statistical analysis of a series of skulls collected by the Australian National Antarctic Research Expedition (private communication) indicates that there are probably subspecific differences between the elephant seal of Heard and Macquarie Islands. It seems possible that there are at least three sub-specifically distinct breeding groups in the Southern Ocean.

Differences in blood serum protein composition occur between elephant seal populations on Heard and Macquarie Islands (Burton, unpublished data).

Bryden (1983) recognised three regional subspecies of the southern elephant seal: *falclandicus* (Falkland Islands and South Georgia); *macquariensis* (Chatham and Macquarie Islands), and *crosetensis* (Iles Crozet, Iles Kerguelen and Heard Island).

Southern elephant seals on Macquarie Island

Early-arriving bull southern elephant seals clumsily haul themselves ashore on the island's shingle beaches in August each year and remain ashore, fasting, until the end of the October–November breeding season. After the bulls come the pregnant cows and juveniles so that, until the end of the following March, the beaches and tussock areas behind them are littered with seals in various stages of growth and development.

Pups, born in September and October, are all weaned by mid-November. After weaning her pup the female is mated (Figure 9.2) and goes to sea for a short time before returning to moult. When all cows have left the island after giving birth the bulls also leave for a time and then return to moult (Figure 9.3). Different age classes moult at different times but, by the end of March, all southern elephant seals have left Macquarie Island to lead a pelagic life during the winter. Some animals haul out on the island for short periods during winter. Yearlings appear to remain in the vicinity of the island for the winter.

Figure 9.2. Bull southern elephant seal, cow and pup, Hasselborough Bay. Bull and cow are mating; pup has black natal fur.

The animals are almost impossible to study during this pelagic phase and very little is known about it. Southern elephant seals are known to be able to stay submerged for more than 30 minutes at depths where they can hunt cephalopods and fish. Stomachs of animals hauled out in winter usually contain cephalopod beaks and fish otoliths.

Southern elephant seal blubber acts as energy source and insulator. Subcutaneous fat is deposited in the epidermis, dermis and hypodermis, the whole tissue mass being referred to as blubber. In southern elephant seals at rest in cold subantarctic waters the insulating effect of the subcutaneous fat layer is adequate to prevent heat loss over a temperature gradient of 36 C° (Bryden, 1964). Blubber of southern elephant seals is thus as effective an insulator as some substances (such as asbestos) used in industry, but is still not as effective an insulator as whale blubber.

When seals which have been ashore leave for sea they appear to head in random directions. Animals from a particular island show no preferred pattern of dispersal as a group. Indications of non-preferential dispersal depend on studies of branded seals which can be recognised individually.

In 1949 the Australian National Antarctic Research Expedition began branding southern elephant seals on both Heard and Macquarie Islands (Carrick & Ingham, 1962a) in order to study their life cycles and movements and fates of individuals. Seals branded on Heard Island were later sighted on Marion Island and in the Vestfold Hills and Windmill Islands areas of Antarctica. Seals from the Windmill Islands have been resighted on both Heard Island and Iles Kerguelen. One animal tagged on Macquarie Island has been sighted in the Windmill Islands. Seals branded on Macquarie Island have been recorded from Chatham Island and vice versa. Despite such known long-distance movements of individuals there is no evidence for interchange between the three main populations (subspecies) in the subantarctic (Carrick, Csordas, Ingham & Keith, 1962). A very much extended marking scheme of seals from each of the three populations would be necessary to provide evidence of genetic interchange, if any, between them. Records of branded or tagged seals suggest that there is no inter-island migration of breeding animals (van Aarde, 1980a). Most long-distance movements are made by immature individuals. In the population based on Macquarie Island, 60% of bulls and 77% of cows breed within 4 km of their birthsite and all individuals breed within 30 km of their birthsite.

Between the early 1800s and 1919 any large southern elephant seal bull hauling out on Macquarie Island in the general region of the Isthmus risked loss of life and blubber to sealing gangs. The same was true throughout the subantarctic.

There has been a general assumption that killing of southern elephant seals for oil led to a dramatic fall in their populations throughout the subantarctic. For instance, Carrick & Ingham (1960) accept that the species was almost exterminated in South America, on South Georgia, Iles Kerguelen, Heard and Macquarie Islands. The idea of a spectacular decline in the species' population on Macquarie Island is probably incorrect, although probably true for elephant seal populations on Heard Island.

Killing of seals for oil on Macquarie Island was confined to areas close to the Isthmus as a matter of simple logistics. The sealers' huts and boiling-

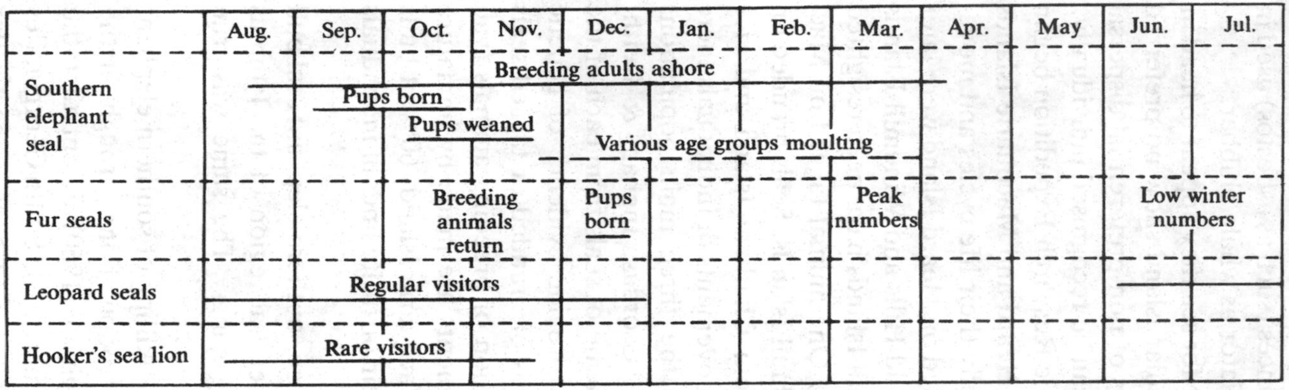

	Aug.	Sep.	Oct.	Nov.	Dec.	Jan.	Feb.	Mar.	Apr.	May	Jun.	Jul.
Southern elephant seal		Pups born	Pups weaned	Breeding adults ashore	Various age groups moulting							
Fur seals			Breeding animals return		Pups born						Low winter numbers	
Leopard seals		Regular visitors						Peak numbers				
Hooker's sea lion		Rare visitors										

Figure 9.3. Timing of life cycle of seals on Macquarie Island. After Anon. (1987)

down works were on the isthmus. As a result, sealers probably had little overall effect on the total island population (Hindell, pers. comm.). Earlier reports support this view.

Cumpston (1968: 139) mentions an abortive attempt by sealers to drive animals from the western to the eastern coast of the isthmus for slaughter:

> Although they had worked the east, north-east and south-west beaches they had only obtained about 10 tons of oil. They had made an attempt to drive the sea elephants from the west to the east side of the island, [possible only in the Isthmus area] but were unsuccessful.

Similarly, Captain Davis, Master of SY *Aurora*, remarked (Cumpston, 1968: 249):

> I may mention that the sea elephant is *not* [Davis' emphasis] on the verge of extinction. On the contrary, the sealers . . . rely entirely for their supply on those animals which come ashore in the vicinity of the digesters.

Ainsworth (Cumpston, 1968: 253) stated:

> It has been stated that these animals are nearly extinct but a visit to Macquarie Island during the breeding season would be enough to convince anybody to the contrary. There are thousands of them, and though about seven hundred are killed during a season, the increase in numbers per year, on Macquarie Island, must be very great.

Cumpston (1968) mentions a total of 1000 tons of elephant seal oil taken from the island in 1826–27. Although 1000 tons of southern elephant seal oil appears to represent a huge kill, the apparently high kill becomes less so when the masses of blubber, muscle, viscera and bone per individual animal are taken into account.

Bryden (1972) dissected a number of southern elephant seals from Macquarie Island into their major components. He found that an 'average' mature male had about 920 kg of blubber and well over 1350 kg of muscle. If one assumes that almost all the blubber is retrievable as oil, a rough calculation shows that about 1200 'average' male southern elephant seals would yield about 1000 tons of oil. One can assume that males rather than females were preferentially sought by oiling gangs because of their higher blubber mass per bodyweight.

Hindell (pers. comm.) believes that, although the sealers worked on Macquarie Island for many years, they may not have made much impression on seal numbers. He points out that Ainsworth (AAE) noted the sealers only worked the Isthmus and a few kilometres on each coast, using only 10% of the total seashore.

If there was, in fact, no 'crash' in populations of southern elephant seals during the heyday of active sealing (continuing until 1964 on South Georgia) suggestions now put forward by various authors that overall southern elephant seal stocks are declining throughout the subantarctic are intriguing. Discussion of this apparent decline appears below.

The breeding season of elephant seals on Macquarie Island

During their time ashore on Macquarie Island during the breeding season, southern elephant seals show complex interactions between sexes and age classes. The animals are gregarious but not what would be described as 'sociable'. Orchestrated chaos among them seems to be the norm. The strong seasonal synchrony of the seals' breeding cycle is reflected in the animals' physiology and behaviour.

The first seals ashore in August are breeding bulls which haul out in preparation for the breeding season which reaches a peak in September–October. Bulls stay ashore fasting until the end of the season. On Macquarie Island the biggest bulls are about 4.5 m long and weigh about 3.5 tonnes. They have a fully developed proboscis (Figure 9.1), are heavily scarred, and, from branding studies conducted on the island, are at least fourteen years old. Carrick, Csordas & Ingham (1962) recognised a hierarchy of bulls from dominant 'beachmasters' to non-breeding 'bachelors'. Older, stronger bulls are the first to arrive and last to leave.

Bull southern elephant seals show seasonal changes in the testis and epididymis (Griffiths, 1984a, b). Testicular mass per tonne bodyweight is lowest (100-150 grams per tonne bodyweight) in May–June. During this period no spermatozoa are produced and the lumina of seminiferous tubules are very narrow and indistinct in histological preparations. As the breeding season (September–October) approaches testicular mass increases to about 300–350 grams per tonne bodyweight. There is a dramatic rise in plasma testosterone levels (Griffiths & Bryden, 1981). The volume of seminiferous tubules increases about three-fold. Sperm first appear in the testis in August but do not reach the epididymis until September and sperm production declines sharply before the end of the breeding season. Successful matings late in the breeding season depend on the sperm stored in the epididymis. Large numbers of sperm are present in the epididymis throughout October and November, become scarce in December and are last recorded in January (Griffiths, 1984a).

Pregnant cows begin to come ashore in the last week of August and gather in harems of different sizes, each harem attended by differing numbers of bulls depending on harem size (Carrick, Csordas & Ingham,

Table 9.1. *Weight gain and loss in southern elephant seal pups pre- and post-weaning.*

Suckling	Week 1	Week 2	Week 3
Daily gain in liveweight (kg)	1.4	5.87	4.71
Post-weaning	Phase 1	Phase 2	
Daily loss in liveweight (kg)	1.52	1.11	

Data from Bryden (1969b).

1962). The number of cows ashore increases steadily from mid-September to mid-October and then decreases abruptly. Cows come ashore about five days before giving birth to a single pup. Twin pups are very rare.

After parturition, cows remain ashore for a further 23 days or so, during which they fast, suckling and then suddenly weaning their pups. During the 23 day suckling period the pup builds up sufficient bodily reserves to tide it over a period of about 35 days post-weaning before it goes to sea. There is a massive transfer of bodyweight from cow to pup during lactation. So profuse is lactation in some cows that skuas are seen drinking milk oozing from mammary glands – a change from their normal role as scavengers responsible for disposal of seal placentas, embryonic membranes and dead pups. Pups suckle frequently and digest the milk rapidly (Bryden, 1968a). During the suckling period pups treble their birth weight, the greater proportion of the weight gain being due to blubber deposition. Rates of weight gain are not constant (Table 9.1).

The pattern of weight gain is typical of aquatic mammals which add body weight preferentially in the order fat–bone–muscle. Non-aquatic mammals follow a different sequence and weight is added as bone–muscle–fat (Bryden, 1969a).

Gales & Burton (1987) used ultrasound to measure blubber thickness of elephant seals on Heard Island. Their calculations give an estimated blubber weight loss of 62.8 kg during lactation in the seals studied; a 48% depletion in the cow's blubber reserves. Total weight loss by a cow during lactation has been calculated to be about 87.5 kg (Gales & Burton, 1987). Total weight gain by a suckling pup is of the order of 84 kg (see Table 9.1). It seems that energy transfer from cow to suckling pup is extraordinarily efficient.

Weaned pups scatter from the harems as the next year's crop of pups is conceived. Cows come into oestrus a few days before or immediately after weaning, usually ovulating eighteen days after giving birth (Griffiths & Bryden, 1981). Their pups are then abandoned and left to their own

Figure 9.4. Group of weaned southern elephant seal pups on beach.

devices (Figure 9.4). Cows and bulls (usually a dominant 'beachmaster') mate and cows go to sea to feed. Weaned pups move to tussock areas and learn to swim, initially paddling in freshwater pools. After six weeks they go to sea for short periods and finally leave the island at eight to ten weeks of age. When the last cows leave so do the bulls (Figure 9.3). Should insemination of a cow have been successful there is a natural delay in implantation of the developing embryo. Copulation occurs in October–November; implantation is delayed until February (Gibbney, 1957).

The description of the breeding period given above conveys little of the bedlam on the beaches during the breeding season. Breeding bulls repel challengers by bluff or fighting for control of harems. A charging bull weighing about 3.5 tonnes and up to 4.5 metres long pays no heed to any cow or pup in its way. Pups seem particularly small in comparison to bulls. A new-born male is only one-eightieth to one-ninetieth the size of an adult bull and the cows are small in comparison to males (Figure 9.2). Breeding animals are not the only seals ashore during the breeding season. Moulting sexually immature animals add to the general confusion.

Quiet periods are spent by the seals in scratching themselves in a remarkably flexible manner for so stout a shape and in mock battles between newly-weaned pups and more serious ones between non-

Figure 9.5. Mature bull southern elephant seals in mild altercation while ashore to moult, Bauer Bay. Tussock vegetation has been severely damaged by seal activities.

breeding males (Figure 9.5). Throwing sand, scooped up by the pectoral flippers and tossed in the air, is another favoured occupation and smaller animals may sink into shallow trenches they themselves have dug.

Control of the southern elephant seal breeding cycle

Timing of the breeding cycle in *Mirounga leonina* appears to be controlled ultimately by changing daylengths between winter and summer. Polar and subpolar regions experience dramatic changes in daylength from winter (very short days) to summer (very short nights). The effect of changing daylength is mediated by the pineal gland which is relatively large in polar animals, including pinnipeds (Griffiths, Seamark & Bryden, 1979). In southern elephant seals, relative weights of the pineal gland are much higher in winter (May–July) than in summer (December–January) (Griffiths & Bryden, 1981). If relative weights of the pineal gland (grams per tonne bodyweight) are expressed as a function of daylength, it is clear that when daylength exceeds about twelve hours pineal gland weight is uniformly low. As days become shorter than twelve hours the gland increases in size as the daylength decreases. This relationship between pineal gland weight and daylength suggests a circadian photosensitive cycle in southern elephant seals.

The pineal gland produces the hormone melatonin which has powerful negative effects on the gonads. Griffiths, Seamark & Bryden (1979) studied plasma melatonin concentration in immature male southern elephant seals during winter and summer. They found a marked circadian cycle in melatonin levels in winter on Macquarie Island (daylength eight hours). Levels of melatonin in the blood showed abrupt changes at onset of light and dark periods, levels at night being significantly higher. In midsummer (daylength nineteen hours) there was no significant difference between daytime and nighttime levels of melatonin. There was no significant difference in mean daylight levels of melatonin between winter and summer.

The pineal system is as yet poorly understood. It has become clear that at least three pineal substances other than melatonin also have potent antigonadotropic effects.

It has been found that newborn elephant seal pups have very large active pineal glands (3.5 g compared to adult 1 g) and that the glands remain large for several weeks. Plasma melatonin levels, however, are markedly reduced a few days after birth (Bryden *et al*, 1986). Bryden (1988) suggests that perhaps the enlarged pineal gland of pups is very important in temperature control in the pup during the first few hours of life. Newborns of many species have stores of brown fat which release a lot of energy for metabolism and temperature control. Tissue that ultrastructurally resembles brown fat in some respects has been observed in newborn elephant seals, but it is uncertain whether or not it is thermogenically active. Melatonin concentrations of 25 picograms per ml blood are common in animals such as adult sheep; elephant seal pups have concentrations up to or even exceeding 6000 picograms per ml blood for the first few days of life. Bryden suggests that this large amount of melatonin is somehow involved in temperature regulation in pups which are completely exposed to a cold environment. After the first meal of milk the pup may then be able to divert fat from the milk into controlling its temperature.

The annual moult on Macquarie Island

Another highly seasonal and synchronised event in the annual cycle of the southern elephant seal is the moult, the precise timing of which varies with age and sex. Immature males and females haul out to moult for 3–4 weeks in November–January. Breeding cows moult in January and February. Breeding bulls moult in February, March and April, older bulls moulting before younger ones. Immature individuals and breeding bulls moult on the beaches. Adult cows remain apart from bulls and occupy tussock or wallow areas (Carrick, Csordas, Ingham & Keith, 1962).

Elephant seal wallows are usually established within tussock grassland just behind the beaches and may be quite extensive (Figure 2.7). Each wallow is usually used by more than one elephant seal (Figure 9.5). Wallows become a foetid mixture of living seals, seal urine, faeces and moulted skin. When wallows occur on deep peats they may become pools of thick mud.

Newborn pups have long, black curly fur (Figure 9.2). This natal fur is lost during the first few weeks as the individual black hairs fall out to reveal a second coat of short, silver-grey hair (Figure 9.4). Timing of the moult in pups is variable. In some it begins as late as 22 days after birth; in others it occurs pre-natally (Carrick, Csordas, Ingham & Keith, 1962). The moulting process in pups bears no resemblance to the annual moult of older animals. A comparable moulting pattern to that in older animals occurs in only one other animal species, the Hawaiian monk seal.

Ling (1965) studied southern elephant seal moulting in detail and described six main stages. The annual moult is rapid and achieved by complete shedding of a cornified epidermis with hair attached. Initially cracks appear in the skin of the axillae, at the base of the tail between the hind flippers and around the eyes and nostrils. Skin with attached hairs begins to slough off and hangs in tatters from the flippers and face which by this stage are almost free of the former coat. The skin then breaks up rapidly on the belly, back and sides and sloughs off. By the time the old coat has been completely shed the seal has a new light grey or blackish grey coat.

Elephant seals have a significant effect on some coastal vegetation (Figures 9.5, 2.7). Post-weaning pups congregate in tussock grassland and do minor damage to vegetation. Subadult and adult seals on the other hand may have significant effects on the vitality of tussocks, particularly during the annual moult. Adult males weigh three tonnes or more and tussocks are killed by seals lying on them or rubbing against them to remove tatters of skin. Some seals travel considerable distances from the coastal beaches over the beach terraces. Stands of *Stilbocarpa polaris* are particularly susceptible to elephant seal damage but regeneration is rapid from underground rhizomes. Urination and defecation, even outside wallows, have significant effects on local soil chemistry. An account of the effects of both seals and birds on coastal vegetation of Macquarie Island is given by Gillham (1961) (see also Chapters 6, 8).

Autumn and winter haul-outs

In the period from mid-February to September some immature seals and occasional pregnant cows come ashore and rest on beaches or in tussock. The haul-out activity is not co-ordinated and may depend on conditions at sea. Cows may come ashore after a miscarriage. The number of animals ashore may drop during calm weather and rise dramatically after and during storms when young seals are likely to come ashore. Old bulls stay at sea the entire winter, often arriving on the island at the start of the breeding season carrying large numbers of goose barnacles.

Comparison with elephant seals from South Georgia

Bryden (1968b) used a growth model to compare elephant seals from the populations on Macquarie Island and the Falkland Islands. He found that mature males on Macquarie Island reached 94% of the bodyweight of Falkland Island animals, but that mature cows were only 60% the size of those from the Falkland Islands. Macquarie Island seals have slower growth rates, later onset of puberty and smaller mature size than those from the Falkland Islands and South Georgia.

Population decline of southern elephant seals?

Early studies of the species on both Macquarie and Heard Islands involved a programme of pup branding to allow fates of individuals to be followed (Carrick & Ingham, 1962a). Branding studies showed that survival of branded pups into their fourth year is about 40% for both sexes. Survival beyond the fourth year decreases markedly in males. Only 25% of males survive to six years and 15% to eight years. Some bulls, however, survive to 20 years or older.

Females have a 40% survival rate to age six (the age of maximum breeding), survival decreasing to 20% by the eighth year. Average breeding females appear to live for between 10 and 15 years (Carrick & Ingham, 1962b). During the 1985 breeding season, Hindell (pers. comm.) sighted two cows branded as pups in 1962, both successfully rearing pups at 23 years of age. Reports by Carrick and his colleagues could cover no more than the period 1950 to 1962. Much more data are now available for Macquarie Island seals but have not yet been thoroughly analysed.

Estimating total populations of elephant seals on Macquarie Island (and other subantarctic islands) is carried out by regular census of animals in particular areas of the island. On Macquarie Island the main area for studying seals is the Isthmus and northern beaches. This area is easily

Table 9.2. *Estimated elephant seal cow numbers ashore on Isthmus at peak of breeding season.*

Year	Estimated number	Year	Estimated number
1950[a]	5307	1962	2507[b]
1951[a]	5166	1963	4534
1952[a]	5528	1971	2834[b]
1954[a]	6360	1972	3885
1956[a]	5765	1975	4388
1957[a]	5035	1978	4412
1958[a]	5861	1980	4236
1959[a]	4777	1982	3000
1960[a]	5732	1984	2895
		1985	2819

[a] Data from Carrick & Ingham (1962b).
[b] Data dubious. Counts may be incomplete.

Table based on Hindell (unpublished data).

accessible from the ANARE station and movement around it is relatively easy.

Carrick & Ingham (1962b) decided that the elephant seal population of Macquarie Island had remained stable over the ten year period 1950–60. They estimated a total population of over 110 000 at the end of the breeding season with about 36 000–37 000 breeding cows and 4000 breeding bulls. More recent studies suggest a decline from these population levels. Hindell (unpublished data) made a census of elephant seals during the 1984–85 breeding season on Macquarie Island and throughout 1985. He counted only 19 000 cows, giving a total estimated pup production of 24 000. These figures represent a substantial decline in numbers of elephant seals on the island when compared with data for October 1960 (Table 9.2).

Elephant seal numbers appear to have declined dramatically elsewhere in some other parts of the subantarctic. Condy (1979) reported a 55.5% reduction in the total population on Marion Island between 1951 and 1976 which included a decline of 69.5% in pups produced in 1976 compared with 1951 (Condy, 1978). Condy thought a number of factors could be responsible for the decline: competition for fish with fishing fleets in the winter feeding grounds of the seals; predation by killer whales (*Orcinus orca*) and competition for local food resources with fur seals whose populations are increasing (Condy, 1978). The decline in the elephant seal population is surprising, particularly as the availability of squid, a major proportion of their diet, has increased following the sharp decline in numbers of whales and the fact that seal culling stopped on Marion Island in 1930. Condy decided that the magnitude of the decline suggests food availability as the major factor involved but does not rule out much-increased predation.

On Heard Island there appears to have been a similar decline in elephant seal numbers since the early 1950s, although census data for the island are missing for a long period during which the base on Heard Island was not occupied. Burton (1986) calculated an overall 60% population decrease over the period. Burton pointed out that in the absence of any major terrestrial changes observable on the islands where seal numbers are decreasing, unknown marine factors must be responsible for the decline. He noted that there have been changes in the atmospheric circulation pattern in the vicinity of Heard Island and that ocean temperature at the island has increased, suggesting that oceanic features will have altered in as yet unknown ways. Burton regards it as difficult to assess the impact of the large fishery on the Kerguelen Shelf on the seal populations which are declining. Macquarie Island temperatures have also risen (Chapter 3).

Elephant seals on Iles Kerguelen represent some 30% of the world population of the species. Census data for 1952, 1958, 1960, 1970 and 1977 show the population to be stable but fluctuating (van Aarde, 1980a). Over the 25 years to 1977, bull and cow components of the population showed different fluctuation patterns. The general trend was for cows to increase in number at a mean rate of 1.4% per year whilst bull numbers declined at a mean rate of 1.9% per year. The decrease in the bull proportion of the population of 8.2% between 1971 and 1977 has led to marked change in the overall sex ratio.

It would appear that there may be factors operating which have differential influences on cow and bull numbers within the total population. Van Aarde (1980a) estimated that the southern elephant seal population of Iles Kerguelen consumed approximately 880 000 tons of squid and 193 000 tons of fish per year. Total fish production on the Kerguelen Shelf (an area of 50 000 km^2) is estimated at 230 000 tons per year, a figure almost comparable to that eaten by southern elephant seals if they fed exclusively in that area. Trawling for fish on the Kerguelen Shelf in 1977 was yielding some 120 000 tons of fish per year – something over 33% higher than the maximum sustainable yield. Such over-fishing is

Table 9.3. *Numbers of southern elephant seals taken annually on South Georgia (1952–65) and oil yield.*

Year	Number taken	Oil yield in barrels[a]
1952–53	6 000	10 807
1953–54	6 000	11 475
1954–55	6 000	11 425
1955–56	6 000	12 068
1956–57	6 000	11 805
1957–58	5 408	11 020
1958–59	5 864	12 476
1959–60	5 787	12 562
1960–61	5 632	12 381
1961–62	4 765	9 666
1962–63	—	—
1963–64	3 998	7 156
1964–65	5 147	9 702
Totals	66 601	132 543

[a] 1 barrel of seal oil = 40 Imperial gallons or 170 kg.

Data from Headland (1984).

bound to have major effects on the Iles Kerguelen elephant seal population.

The largest population of southern elephant seals in the world, that on South Georgia, remains stable (Headland, 1984). Sealing on South Georgia has been quite firmly controlled and regulated since 1909 and the industry, which ceased only in 1965 when whaling also came to an end on the island, was based on the assumption of maintaining a sustained yield. Regulations included a closed season for hunting during the breeding season, a four-sector rotation plan one sector of which was untouched each year, and limits on the sex, age and numbers of elephant seals killed. New regulations were introduced in 1952 which led to a sharp rise in the catch per unit effort and higher yields of oil (Table 9.3). There was also an increase in the average age of the catch. The South Georgian population, at about 300 000 individuals, has remained stable, even though some 6000 bulls were killed each year.

Fur seals

The stimulus for much early exploration of high southern latitudes was the quest for fur seal pelts, an extension of a long-established industry in the northern hemisphere. Sealers quickly followed in the wake of discoverers (or were discoverers themselves) and a subantarctic fur seal industry was well-established by the mid-eighteenth century. No attempts were made to conserve stocks. Seals of all ages and sexes were taken and the industry thus became self-limiting when fur seal stocks became decimated. Fur seals were wiped out in some localities, one of them being Macquarie Island.

The original fur seal population of Macquarie Island was obliterated within a decade of its discovery. By April 1811, some 56 974 skins had been taken from the island. By the end of 1812, the number taken stood at 120 000 and had risen to 180 000 by 1813 (Csordas & Ingham, 1965). It has usually been assumed that the New Zealand fur seal (*Arctocephalus forsteri*) was the original fur seal on the island but Falla (1962) has suggested that it may have been another and more vulnerable species, possibly the Antarctic fur seal (*A. gazella*). It may also have been a mixture of species as at present. All ideas about precisely which species was or were present in the original population have to be assumptions. There are no known specimens on which to base a decision. Sealers called them 'upland seals' and their coats appear to have been more luxuriant than those of New Zealand fur seals. The original population may even represent a now-extinct species.

Fur seals were not reported by members of the AAE based on the island during 1911–14 although, at that time, sealers working the elephant seal population reported occasional sightings of them (Mawson, 1943; Csordas, 1958). When the permanent station was established on the Isthmus in 1948 a small colony of non-breeding fur seals was found near North Head. It was thought that these animals may have come from the Auckland or Campbell Islands. During 1949–50, Gwynn (1953b) counted 174 individual fur seals. By 1957, there were 298. In 1954, a small colony of fur seals was found at Hurd Point and another at Handspike Point in 1955. When these colonies became established is not known.

Law & Burstall (1956) described the population as composed of 'mainly immature seals and pups a year old, a few of the latter being still accompanied by their mothers'. The first possible breeding of fur seals on Macquarie Island this century was recorded in March 1955 (Csordas, 1958; Csordas & Ingham, 1965).

It is now known that the island is being recolonised by three species of

Table 9.4. *Distinguishing features of the three species of fur seals found on Macquarie Island.*

*Subantarctic fur seal (*Arctocephalus tropicalis *)*
Adult male: Cream chest and face. Utters repeated deep growl when approached. Adult female: Fur colour variable.[a] Pups: Black, including muzzle. Hybridises with Antarctic fur seal.

*Antarctic fur seal (*Arctocephalus gazella*)*
Adult male: Uniform grey to dark brown. Stands ground and whimpers when approached. Adult female: Fur colour variable.[a] Pups: Broad flat head, blunt snout, fur grizzled. Muzzle whitish. Hybridises with subantarctic fur seal.

*New Zealand fur seal (*Arctocephalus forsteri *)*
Adult male: Fur colour similar to Antarctic fur seal, but fur coarser and browner. Snout longer and paler than Antarctic fur seal. Growl and then flee when approached. Adult female: uniformly brown. Pups: Rounded head profile, white muzzle. Appears not to breed on Macquarie Island.

[a] Fur of cows of subantarctic and Antarctic fur seals similar, making it difficult to separate species. Species separable on basis of snout shape, size of external ears and vocalisation.

Data from Shaughnessy & Shaughnessy (1987a).

Figure 9.6. Fur seals at Secluded Beach. *Arctocephalus tropicalis* bull (right) with *A. tropicalis* or *A. gazella* cow (left). Photograph G. Little.

fur seal (Table 9.4). The New Zealand fur seal, which is known to have been present on the island for almost 40 years has not succeeded in forming a breeding population (Shaughnessy & Shaughnessy, 1987a, b). In 1986–87, most territorial bulls were *Arctocephalus tropicalis* and most cows *A. gazella* (Shaughnessy & Shaughnessy, 1987a) (Figure 9.6). Shaughnessy & Shaughnessy (1987a) discuss problems in identification of the three species present.

Arctocephalus tropicalis and *A. gazella* occur together on Iles Crozet and on Marion Island. In 1986-87 on Macquarie Island five breeding territories with more than one cow contained adults of both species. Interspecific hybridisation appears to occur. Six of 37 pups in the colonies showed characteristics of both species (Shaughnessy & Shaughnessy, 1987a).

Fur seals on Macquarie Island are found on rocky beaches and among coastal rocks. They may lie on *Colobanthus muscoides* cushions but do not yet cause significant vegetation damage. Any increased nutrients are probably washed away by rain. They supplement their fish and squid diet with royal, gentoo and possibly rockhopper penguins. An increasing population of fur seals on the island may increase predation pressure on penguins.

In February 1982, 1245 fur seals were counted on the island and twenty pups were counted during the 1982–83 summer. Four large bull subantarctic fur seals (*Arctocephalus tropicalis*), distinguishable from all other fur seals on the island because of their markings and colour, were noted (Fletcher & Shaughnessy, 1984). Females in the breeding colonies were smaller in size and differed in colour from all bulls of either *A. tropicalis* or *A. forsteri*. The females were initially thought to be *A. tropicalis*. A few have subsequently been identified as Antarctic fur seals (*A. gazella*).

While the majority of bulls on Macquarie Island are non-breeding *Arctocephalus forsteri*, the two groups of breeding animals consist of *A. tropicalis* bulls and either *A. tropicalis* or *A. gazella* cows. In the 1981–82 breeding season, there were four breeding bulls, fifteen females and fifteen pups. During the following breeding season, the composition of the breeding colonies was five *A. tropicalis* bulls, two subadult *A. tropicalis* bulls, thirty cows (either *A. tropicalis* or *A. gazella*) and twenty pups. Weaning patterns and behaviour are consistent with those seen in *A. tropicalis* and *A. gazella* elsewhere (Fletcher & Shaughnessy, 1984). Pup numbers are increasing.

Although non-breeding fur seals were first seen on the island in 1950 and the first breeding animals reported in 1955, it is possible that the island was

Table 9.5. *Self-introduced mammals on Macquarie Island.*

Species	Year of introduction	Comments
Black rat[a]	? Very early after discovery	Probably off-loaded with stores[b]
House mouse[a]	? Very early after discovery	Probably off-loaded with stores[c]
? Cat[a]	? Very early after discovery, prior to 1820	Bellingshausen mentions wild cats (1820)[d]

[a] Breeding population established on island (1987).
[b] Earliest historical account from 1908 (Cumpston, 1968: 219).
[c] Earliest historical account from 1910 (Cumpston, 1968: 209). Both black (ship) rats and house mice are commensal with humans and would presumably have been present on early sealing ships.
[d] Cats were commonly present on sealing ships to keep down rats (Cumpston, 1968: 46).

colonised or recolonised by both *Arctocephalus forsteri* and *A. tropicalis* bulls at about the same time. The breeding females of either *A. tropicalis* or *A. gazella* may have arrived shortly after.

Fur seal numbers have increased dramatically throughout the subantarctic. Budd (1970) and Budd & Downes (1969) discussed population increases in *Arctocephalus gazella* on Heard Island. Between 1952 and 1975 there was a 10.5% annual population increase of *A. tropicalis* on Marion Island where both *A. tropicalis* and *A. gazella* breed (Condy, 1978). Increases in numbers of *A. gazella* on South Georgia have been documented (Bonner, 1964, 1968; Payne, 1977). Range extensions of *A. gazella* have been noted to many islands south of the Antarctic convergence (Bonner, 1976). Populations of *A. tropicalis* are also increasing. It is possible that increased food availability – chiefly krill for *A. gazella* (Laws, 1977) and fish and cephalopods for *A. tropicalis* (Rand, 1956; Condy, 1978) – may have led to the increase in numbers. The decline in elephant seals noted over the same areas may be related to direct competition between elephant seals and fur seals for food.

Non-indigenous mammals

The recorded history of the addition of alien species to Macquarie Island's ecosystem (and the extinction of some indigenous species as a result) begins with the self-introduction in 1810 of *Homo sapiens*. Wherever human populations are established, even on a very temporary basis, other species are likely to follow. Numerous non-indigenous animal species have been taken, not always knowingly, to Macquarie Island by

Table 9.6. *Species deliberately introduced to Macquarie Island.*

Species	Year of introduction	Comments
? Cats		See note [d] of Table 9.5.
Dogs	Prior to 1820	See [b] below
Rabbits[a]	1878 or 1879	Deliberately introduced as food for sealers (Cumpston, 1968: 97, 118, 119, 159)
Donkeys	1878	To be used for carrying blubber at which they were unsuccessful (Cumpston, 1968: 113, 115)
Horses	1917	To be used to pull blubber sledges (Cumpston, 1968: 299, 307)
Cattle	1954	Introduced by ANARE as food source. Line established on island for several years died out in 1960 (Csordas, 1985)
Sheep	? 1820	See *c* below
	1911	Australasian Antarctic Expedition landed 12 sheep as food for island party
	1913	25 sheep landed to reinforce stock (Cumpston, 1968: 265)
	1948	ANARE landed 33 sheep as source of food (King, 1985)
Goats	1878	First goats introduced were for food (Cumpston 1986: 115–16, 118, 159).
	1948	ANARE landed 12 goats which became feral. Last goats killed 1957 (Csordas, 1985)
Pigs	1878	(Cumpston, 1968: 115–16)
	1953	ANARE for food (Csordas, 1985)
	Prior to 1961	(McLeod, 1986)

[a] Breeding population established on island.
[b] Bellingshausen (1820, in Cumpston, 1968) mentioned wild dogs on Macquarie. There is no other historical evidence to support his statement. All dogs taken to the island were pets, often used for hunting. Dogs were valuable to the sealers as hunters to supplement an otherwise monotonous diet.
[c] Sheep were suggested as the basis for an industry on Macquarie Island by Mawson (1919); see Cumpston (1968: 279, 318, 322). The sheep mentioned by Bellingshausen were probably from his own ships.

parties visiting the island, and three mammalian species, cats, mice and rats, have become accidentally established there in breeding populations. The known history of the introduction of alien mammals to the island's biota is summarised in Tables 9.5 and 9.6.

Accidental introductions

Cats

Cats (*Felis catus*) on the island today are presumably descended from ships' cats (the old-fashioned rat trap) or from pet cats taken ashore with early sealing parties. Domestic cats have been used at the permanent ANARE station in their traditional role as catchers of rodents and rabbits. Cats kept at the base are now desexed and do not contribute to the overall problem of the island's feral cat population.

It seems clear that feral cats had become established on Macquarie Island soon after its discovery. Today there is a concerted campaign to rid the island of cats, a campaign first suggested by Douglas Mawson who recognised, as early as 1919, the danger posed to some indigenous birds by feral cats. A campaign to eradicate cats from Macquarie Island may well succeed. Trying to rid the island of rodents would be a different problem altogether.

Feral cats, found over about 65% of the island, are common in all habitats except plateau feldmark (Jones, 1977). Most live in herbfield or tussock grassland (Brothers, Skira & Copson, 1985). Of 246 cats collected between 1976 and 1981, 74% were tabby, 26% orange and 2% black. The breeding season extends from October to March with a peak in November–December. Some females may produce two litters a year. In 1985, the total island cat population was estimated at 169–265. The home range of one male covered 41 hectares in summer, but it did not remain in it during winter when food supplies became scarcer (Brothers, Skira & Copson, 1985).

Analysis of cat scats showed rabbits to be cats' main dietary item, with prions and petrels next in importance (Table 9.7). Rabbit kittens to 300 grams were most frequently taken (Table 9.8). The scats analysed are not representative of any particular season as there is no way of telling when they were dropped (Jones, 1977). Analysis of gut contents of an additional 41 cats during winter showed a change in diet. More wekas were taken and more time was spent scavenging dead southern elephant seals and penguins (Jones, 1984). Jones pointed out that the total amount of food available during the winter, when burrow-nesting petrels are absent from

Table 9.7. *Food items detected in cat scats.*

Food item	Number of scats	Frequency (%)
Rabbit	619	81.9%
Prion	220	29.1%
White-headed petrel	120	15.9%
Mouse	33	4.4%
Penguin	25	3.3%
Rat	20	2.6%
Weka	15	2.0%

Data from Jones (1984).

Table 9.8. *Weights of rabbits taken by cats.*

Weight of rabbit	Frequency in cat diet
200–300 grams	58%
300–600 grams	23%
600–1300 grams	8%
More than 1300 grams	11%

Data from Jones (1984).

the island, is probably the major factor limiting the cat population. Jones (1984) estimated the adult cat population in 1974 at 250–500 adults, a much higher population than reported since.

On Macquarie Island, shooting is used as the primary control measure for cats. During 1987, 161 cats were shot over a nine month period and it is reported that cat numbers may now be quite low. More than 100 cats have been shot annually for some years. It may be that kill numbers approximate annual successful litter production.

On Marion Island, the development of a feral cat population has been documented by Watkins & Cooper (1986). Cats on Marion Island appear to be descended from four cats introduced to the meteorological station for rodent control in 1949. The first feral cat was seen in 1951, and 16 years after their introduction cats were widely distributed over the island up to altitudes of 450 m. Feral cats appear to have taken a heavy toll on Marion Island bird life (e.g. an estimated 455 000 birds killed by cats in 1975). In 1977, feline panleucopaenia virus was introduced to Marion Island. If estimates of the numbers of cats present on the island at various times are correct, there followed a massive drop in the population as well as a reduction in fecundity and changes in the age structure of the population. Intensive night shooting is now undertaken during the breeding season (synchronised in Marion Island female cats) as a secondary control measure aimed at eliminating all female cats from the island.

Rats

In the early sealing days, with small ships carrying men and provisions to Macquarie Island, every opportunity existed for ship or black rats (*Rattus rattus*) and the house mouse (*Mus musculus*), both of which live essentially as commensal species with humans, to reach the island.

Macquarie Island must be one of the few places in the world where the black rat (*Rattus rattus*) presently lives free from competition with the brown rat (*Rattus norvegicus*) which has largely displaced it elsewhere. Rats reached the island at an unknown date but have plagued people on the island ever since.

A member of a sealing party in 1908 recorded in his diary:

> both [huts] were over-run with large rats, which were at first inclined to dispute possession. The newcomers found themselves under attack and, when they protested, the rats merely moved back out of reach. They also found their way into the mattresses, which consisted of two sacks sewn together and partially filled with chaff. If a man rolled onto one it would squeal and it was then necessary to cut the bag open and tip the contents onto the floor, whereupon the rats would sort themselves out and the chaff could be shovelled back into the bag.
> Despite the rats and the rain the men slept soundly.
>   (Cumpston, 1968: 219)

The Macquarie Island party of the AAE also suffered from the activities of rats. In 1913:

> They [Blake and Hamilton] moved on to Lusitania Bay on 19 September, where it was found that rats had destroyed some of Hamilton's skins at the hut . . .
> On 30 October Blake made a special trip to Sandy Bay to bring back some geological specimens and other things he had left there. He was surprised to find that the Sandy Bay hut had been burnt to the ground . . . It was thought that the rats had started the fire from wax matches which had been left lying on a small shelf.   (Cumpston, 1968: 268)

Rats still plague expeditioners using the field huts on the island, as shown by an extract from the log book of the Lusitania Bay hut. The entry, dated 25 June 1986, reads (in part):

> It was wonderful to crawl through the Luci hut entrance hole and see the
> wondrous damage Ratus Ratus [sic] had caused in his MIDWINTER
> spree. He's been in every packet, every tin and left cute Ratus droppings
> everywhere! Poor dear sweet Ratus had even lovingly attacked 2 sleeping
> bags and distributed the white soft fluffy dacron filling to cosy hideaways
> in the hut . . . Hopefully we'll consume enough Fielders [rum] so dear
> Ratus Ratus wont keep us awake tonight . . .

Despite the problems they cause, the biology of rats on Macquarie
Island has not been intensively studied and there is no systematic attempt
being made to eradicate them from the island. Copson (1986) investigated
the diet of rats by analysing stomach samples. He found that rats ate
mainly plant matter, with *Poa* species common in spring and summer but
almost absent in autumn and winter. Fruits and stems of *Stilbocarpa
polaris* are also eaten, along with invertebrates and some vertebrate
material the nature of which was unspecified. The extent to which rats
threaten burrow-nesting birds is not known.

Mice

No one knows when house mice became established on the island.
They were certainly present when, in 1901, members of Scott's Antarctic
expedition went ashore from *Discovery*. Wilson, ornithologist with the
expedition, attributed partial destruction of a collection of prepared bird
skins in one of the huts to mice: 'in many cases the legs, feet and wings [of
the bird skins] had been eaten by mice' (Cumpston, 1968: 209). Mice,
small and secretive, may have been overlooked earlier.

Mice appear to live all around the island up to altitudes of 300 m.
Favoured habitats include stands of tall tussock grass, *Stilbocarpa polaris*
and grassland dominated by species of *Agrostis, Festuca, Luzula* and
Deschampsia. Pye (1984) found their diet consisted mainly of *Stilbocarpa
polaris* seeds and other plant material, with a garnish of invertebrates.
Copson (1986) found that invertebrates, particularly moth larvae and
spiders, formed the main dietary item in the mice he studied. Plant
material appeared to be of little importance. Further studies are needed to
resolve the apparent differences between these reports.

Berry & Jakobson (1975) included mice from Macquarie Island in a
general study of wild-living mice to determine whether mice living under
harsh environmental conditions have any characteristics not found in mice
living under less extreme conditions. Berry & Peters (1975) studied
Macquarie Island mice to determine the effect of life in an equable but

stressful environment. They concluded that the animals were not remark-
able in any traits although they were fairly large and had relatively short
tails. Berry and Peters reported year-round breeding in Macquarie Island
mice. Pye (1984) pointed out that reproduction in free-living populations
of house mice can be either seasonal or non-seasonal. Pye systematically
collected mice on Macquarie Island throughout a year and found that the
evidence points to a seasonal breeding pattern. Twenty-nine female mice
trapped during winter showed no signs of reproductive activity. There
were gross changes in uterine and ovarian structure with a complete
absence of corpora lutea in some of the animals and atrophied corpora
lutea in others.

Deliberate introductions

Sheep

In 1911, SY *Aurora*, carrying the AAE southwards, put ashore
twelve sheep which were pastured on Wireless Hill. They thrived, grew fat
and one was killed each month to feed the Macquarie Island party in the
days before stations were equipped with freezers. In 1913, the sheep
population was replenished by the *Tutanekai* which put ashore 'Two more
boat loads of stores, including 25 sheep and 22 cases of coal . . .'
(Cumpston, 1968: 265).

A. Tulloch, who staffed the meteorological station until the end of 1915,
had a flock of sheep consisting of three Shropshire Crossbreds, three
Comebacks, four Lincoln ewes and two Shropshire rams, as well as
fourteen killing sheep. These animals withstood the climate and bore
lambs which thrived. Tulloch reported that he considered the Comebacks
and Shropshires best suited to the island and considered the possibility of
establishing sheep farming on the island. He wrote:

> The sheep ate the tussock down in many places and appeared to thrive as
> well on the tussock as on the Maori cabbage [*Stilbocarpa polaris*].
> The island slopes, amounting to at least one fourth of the whole island,
> are densely covered with tussock and Maori cabbage, totalling, I would
> say about 6000 acres. The land should carry three sheep to the acre, giving
> the island a carrying capacity of 18,000 sheep. The sheep would in no way
> affect the sea elephants or the penguin oil industry, in fact they could be
> run in conjunction. (Cumpston, 1968: 279)

The then Governor of Tasmania, Sir William Allardyce, had lived for a
decade in the sheep-farming and sealing community of the Falkland
Islands and proposed, in 1920, to both the Tasmanian Premier and the

Governor-General, that a small flock of sheep be established on Macquarie Island. The necessary personnel with a knowledge of both sheep and seals would have been recruited from the Falkland Islands (Cumpston, 1968: 318).

Even Mawson, so anxious to end unfettered exploitation of Macquarie Island's birds and animals, did not advocate the complete protection afforded them today. He visualised a permanent weather station with sheep and reindeer to provide for the people staffing it, recognised that some bull elephant seals could still be harvested for oil (an idea which later proved feasible on a sustained basis on South Georgia) and thought of the possibility of an industry based on dried penguin eggs (Mawson, 1922).

When Tulloch shut down the meteorological station at the end of 1915 he left behind six ewes, three lambs and two rams. By 1918 there were about four sheep left. A couple of them dived over a cliff rather than be caught. When the sealing party left the island, one sheep was left to roam at will after having the wool cut from its eyes (Cumpston, 1968).

Rabbits

The European rabbit (*Oryctolagus cuniculus*) was taken to Macquarie Island with the express intention of establishing breeding populations as a food source for members of sealing gangs. The introduction was highly successful and rabbits now cause problems with the island's ecological balance.

Cşordas (1985) places the introduction of rabbits at some time in the 1860s or 1870s. In April 1891, William Elder, in evidence to an inquiry into the loss of the *Kakanui*, a small vessel plying between New Zealand, Macquarie Island and New Zealand's subantarctic islands, stated:

> I procured and sent to the islands some French rabbits in 1879 or 1880 and I know there are millions of rabbits there now. There were none before I sent them down . . .
> Mellish was working for me when I sent the rabbits down.
> (Cumpston, 1968: 118)

Elder's statement is backed up by the report of a party which had been on Macquarie Island in 1877:

> There were no rabbits at the time of our stay on the island. They were, however, imported by Elder and Co, who succeeded our party, and since then they have increased enormously.   (Cumpston, 1968: 97)

On 12 December 1890, a New Zealand newspaper, the *Dunedin Evening Star*, published a letter from a person signing himself 'E.R.'.

Cumpston (1968) suggests that 'E.R.' was Edward Risk. 'E.R.' stated:

> There were no rabbits when we went down; in fact our party liberated the
> first on the island, and protected them as much as possible while we were
> there, and had the satisfaction of seeing numbers running around before
> we left . . . We let them go at what we called then North-East Harbor
> [sic] . . .   (Cumpston, 1968: 119)

If 'E.R.' was indeed Edward Risk, he would have been a crew member
of, or sealer carried on, the schooner *Jessie Nicol* which was sent to
Macquarie Island by Cormack, Elder and Company and left a party there
in 1878. Elder's mention of Henry Charles Mellish in the context of rabbits
is also significant. Cumpston (in the index to his 1968 work) notes Mellish
as employed by Elder in 1879. In April 1890, a party under Mellish was on
Macquarie and reported 'There was any quantity of rabbits to be had on
the island' (Cumpston, 1968: 159). If there was 'any quantity' of rabbits
they must have been introduced some time before 1890 to allow for
significant population growth. The historical evidence available points to
either 1878 or 1879 as the year of introduction of the European rabbit to
the island.

By 1974, rabbits were common on 50% of the island area, living mainly
in herbfield areas (Copson, Brothers & Skira, 1981) and retaining the basic
behavioural pattern of domestic rabbits in their tendency towards diurnal
activity. During winter they tend to stay underground with only short
surface forays, but spend over 90% of the day above ground in summer.
Numbers on the island peaked in the summer of 1977–78. Numbers have
fluctuated and declined since then. In 1956 and 1965–66 the rabbit
populations were estimated at 50 000 and 150 000 respectively. In 1974, an
estimate of 50 000 was made, and in 1977–78, 150 000. It would seem that
large-scale annual fluctuations are the result of weather changes.
Extended dry periods would lead to an increase in numbers; wet periods
cause large-scale mortality, particularly of kittens in flooded burrows.
Predation pressure by cats and skuas probably remains fairly constant.

Rabbit activities can have dramatic effects on vegetation. Heavy grazing
pressure may devastate local areas. *Pleurophyllum hookeri* and
Stilbocarpa polaris, in particular, may be grazed down to stumps (Figure
9.7). *Azorella macquariensis* cushions, although evidently not eaten by
rabbits, are often damaged by their scratching (Copson, 1984).

The role of rabbits in causing accelerated erosion on the island is
discussed in Chapters 5 and 6. Although selective rabbit grazing has been
important in changing the floristic composition and structure of grassland

Figure 9.7. Rosettes of *Pleurophyllum hookeri*. (a) Ungrazed. (b) Leaves have been grazed by rabbits, leaving leaf bases at ground level.

and herbfield vegetation, the role of rabbits in causing erosion of grassland sites is difficult to assess and may have earlier been overestimated (Selkirk, Costin, Seppelt & Scott, 1983).

There is now an active and effective programme to reduce the rabbit population on the island using myxoma virus.

10

Microbiology, parasitology and terrestrial arthropods

Macquarie Island, in common with any land surface with extensive plant cover, supports a rich and varied assemblage of soil and plant microorganisms. The composition and ecological roles of the island's microflora and microfauna, both free-living and parasitic, are as yet poorly known.

Soil organisms

Bacteria, algae and nematodes

Bunt (1954a,b,c) described results from a broad microbiological survey of Macquarie Island soils. Information from this study, conducted in 1951–52, still provides the most up-to-date data on soil nematodes and soil algae on the island. Bunt reported the presence of 'iron' bacteria (*Crenothrix* sp.) in situations constantly saturated with drainage water, the bacteria precipitating iron hydroxides from the water. We have observed iron-cemented wind-blown sands between Bauer Bay and Sandy Bay and iron-organic layers beneath peat. In both cases biological activity was probably involved in mobilisation and precipitation of the iron. Bunt described filamentous sulphur bacteria (*Thiothrix* spp.) in slowly moving water in herbfields, and purple photosynthetic bacteria (possibly *Rhodopseudomonas* spp.) from a mire soil in herbfield.

Bunt (1954b) recorded a total of 126 diatom species from the island, in soils, lakes and ponds, fossil deposits such as deep peats, and other sediments. His taxonomic work on diatoms is probably in need of revision, using more modern optical and electron microscopic techniques, but his general conclusions are still valid. He detected differences between diatom populations in soil, in lakes and ponds and in fossil deposits and commented on the possible significance of diatoms in overall soil ecology on the island. He pointed out that rotifers (in culture) can subsist on soil

188

Table 10.1. *Number of species of each major fungal group on Macquarie Island.*

Basidiomycetes	at least 41 species
Ascomycetes	at least 27 species
Imperfect fungi	
Coelomycetes	at least 15 species
Hyphomycetes	at least 41 species
Lower fungi	
Zygmomycetes	at least 6 species

Data from Kerry (1984).

algae and that some free-living nematodes are algal feeders, as are Collembola.

Bunt (1965) provided estimates of numbers of bacteria and algae per gram of soil. He found considerable microfloral (algal and fungal) variation between samples, with fungi making up only 0.5–1.5% of the total soil microflora in most samples. He listed 27 fungal genera (including plant parasites, general saprophytes on litter, and soil fungi) from the island. The numbers of soil-inhabiting fungi are comparable to those in soils from more temperate regions. *Penicillium* species appear to dominate the fungal flora.

Kerry (1984) continued Bunt's work on fungi, making a general survey of fungi on plants and litter of *Poa foliosa*, *Stilbocarpa polaris* and *Pleurophyllum hookeri*, and following the pattern of fungal succession (Kerry & Weste, 1985). It is now known that all four major fungal groups are present on the island (see Table 10.1 and Appendix 4).

In experiments on the effects of temperature on growth of various fungal species from Macquarie Island, Kerry (1984) found that many leaf-colonising fungi grew faster at 4 °C than other fungi, although they had highest growth rates at 15 °C–20 °C and showed little or no growth at temperatures in excess of 25 °C. Kerry's results suggest that development of the island's leaf microflora has involved selection for species and/or strains able to grow at low temperatures even though, at the temperature prevailing on the island (about 5 °C; annual average temperature 1949–86 = 4.8 °C), they grow at only 10–20% of their maximum rate.

Free-living nematodes, a major component of the soil ecosystem, were studied by Bunt (1954a). He collected soil from a number of localities on the island and concluded that number and types of bacteria and algae in each sample were of major importance to nematode populations present. Most nematode species identified in the soil samples are bacterial and/or algal feeders, and the same is probably true of most nematode species which have not yet been identified.

Bunt (1954a) demonstrated a general increase in soil nematode populations between August and February in a number of samples from different localities. The slight differences between winter and summer on Macquarie Island are apparently sufficient to influence reproductive capacity of nematode populations, either directly or by influencing populations of soil bacteria and algae. Bunt made no attempt to ascribe the increases to any particular cause and appears to have ignored the major variable in climate on Macquarie Island, the substantial difference in length of daylight between winter and summer, as a possible cause of the differences. The difference in daylength between winter and summer is likely to have major effects on small photosynthetic organisms. Whatever the cause or causes, there are differences in nematode populations between summer and winter.

Nematode populations in penguin colonies show different trends depending on the particular species of penguin present (Bunt, 1954a). Assuming that nematode numbers in tall tussock grassland bordering lowland herbfield are reasonably similar from area to area around the island, Bunt studied nematodes in a gentoo penguin colony established in September in previously undisturbed tall tussock bordering lowland herbfield. The colony contained several hundred birds which damaged vegetation and laid bare the peaty soil. He found a twenty-fold increase in nematode numbers on the site when compared with similar sites undisturbed by gentoo penguins. The increase in nematode numbers is probably caused by multiplication of soil bacteria in response to input of penguin excrement. Bunt suggested that localised increases in soil temperature directly below nesting birds may also be a factor leading to increase in nematode numbers in gentoo penguin colonies.

A study of a rockhopper penguin colony showed an opposite trend. In August, when the colony was unoccupied, the nematode population was $124 \times 10^6 \mathrm{m}^{-2}$ (based on sampling the surface 15 cm of soil). Numbers had dropped to $32 \times 10^6 \mathrm{m}^{-2}$ at the end of September, the start of the breeding season. By the following February, when the penguins were ready to leave, the nematode population had fallen to $9 \times 10^6 \mathrm{m}^{-2}$. Apparently conditions are inimical to nematodes when the birds are present. Total bacterial counts from soil samples in the colony remained fairly constant at $2200 \times 10^6 \mathrm{g}^{-1}$ in August and $2800 \times 10^6 \mathrm{g}^{-1}$ in February. Some factor other than food supply must limit nematode populations in rockhopper penguin colonies during the penguins' breeding season.

The opposing trends shown by nematode populations in colonies of the two penguin species are interesting. The cause is unknown and shows how much detailed work has yet to be done on soil microflora and microfauna on Macquarie Island before even a preliminary understanding of population dynamics within soil ecosystems on the island can be established.

Bunt (1954a) found that a number of algae and bacteria in Macquarie Island soils was much higher in February than in August, summer numbers being 1.5 to 20 times the numbers present in winter. He decided that this increase in the soil microflora was at least partly responsible for differences between summer and winter populations of nematodes but pointed out that no clear relationship existed between total nematode populations and total bacterial and algal numbers in the samples studied. He suggested that selective feeding by nematodes on particular elements of the soil microflora could explain the lack of a clear correlation between nematode populations and total soil microflora.

Soil arthropods

Free-living and terrestrial arthropods of Macquarie Island have been little studied except for taxonomic purposes and compilation of lists of species present (Watson, 1967; Gressitt *et al.*, 1962; Greenslade & Wise, 1986) Ecological studies have largely been confined to arthropods parasitic on birds and seals (see below) and a detailed study of the copepod *Pseudoboeckella brevicaudata* (Chapter 7).

Greenslade (1987) carried out a survey of island Collembola which are normally abundant on subantarctic islands, outnumbered in the soil arthropod fauna only by mites. Greenslade pointed out that Collembola must make a very large contribution to nutrient cycling in soils on Macquarie Island in the absence of other detritivores such as myriapods, isopods, amphipods and numerous insect larvae. She found microarthropod densities in soil samples of 100 000–250 000 m^{-2}. Tall tussock grassland has the most diverse microarthropod fauna. Low densities of microarthropods occurred in feldmark samples studied.

Bacteria of birds and mammals

Bunt (1955) studied the faecal floras of southern elephant seals and 13 bird species on Macquarie Island. The primitive conditions under which he worked during the Australian National Antarctic Research Expedition of 1951–52 caused great problems in maintaining cultures and many isolates were lost. Bunt isolated *Escherichia coli* from over half the species studied and found no other bacterial types in faeces of southern elephant seals or king penguins. He identified both *E. coli* Type 1 and *E. coli* Type 2. A species of *Bacillus* was isolated from both royal and gentoo penguins. Diving petrels had the most diverse faecal flora. Attempts at isolating bacteria from faeces of dove prions and white-headed petrels failed. Bunt also examined wound pus and nasal mucus from bull southern elephant seals. The pus contained several types of bacteria which were not fully identified. He found that infections of the nasal passages of southern elephant seals are prevalent.

Virology on Macquarie Island

Viruses, ticks and penguins

Since the tick *Ixodes uriae* is known to carry viruses and may transmit them both within and between seabird species, and considering that birds and seals from Macquarie Island and other subantarctic areas carry other ectoparasites and endoparasites, considerable work on ticks of seabirds and the viruses they carry has been done on the island.

Microbiological studies on Macquarie Island have lately emphasised virological studies. It has been realised that viruses, particularly if newly introduced to 'innocent' previously isolated populations, may be expected to have major effects on those populations. The most comprehensive studies of naturally occurring viruses on Macquarie Island are those of Morgan and his co-workers.

Morgan, Caple, Westbury & Campbell (1978) took cloacal and tracheal swabs from royal, king, gentoo and rockhopper penguins, northern giant petrels, skuas and cormorants. Swabs were examined for antibodies to known and detectable viruses. Particular emphasis was placed on a search for Newcastle Disease virus and Influenza A virus. Newcastle Disease virus causes respiratory tract disease in domestic poultry (particularly fowls and turkeys) and may even affect poultry workers and laboratory personnel. The connection between avian reservoirs of Influenza A virus and outbreaks of Influenza A in human populations has now been well established. Blood serum samples for virological examination were taken from the four species of penguin on the island and from skuas.

Serological tests show that 6% of royal penguins sampled had antibodies to Newcastle Disease virus. It is not known where or how the penguins come in contact with the virus, but they appear to contract it while away from the island on their annual post-breeding migration. The birds have shed any active virus particles before they return to the island. It appears, from the ages of birds sampled, that they first contact the virus about the time they reach breeding status.

The viral strain involved is identical to that found in Australian domestic poultry. It is a mild strain which has been used to vaccinate Australian poultry against fulminating Newcastle Disease. Thus there may well be a degree of immunity to Newcastle Disease among royal and other penguins on Macquarie Island. Perhaps the viral strain was originally introduced to the island by the Australasian Antarctic Expedition 1911–14 or by early ANARE expeditions which included poultry among their provisions.

Royal penguins breed only on Macquarie Island. They regularly visit Heard Island, however, and have been recorded on various parts of the coastlines of Tasmania, Victoria and South Australia. The fact that royal penguins on Macquarie Island have antibodies to Newcastle Disease virus implies a need for application of strict quarantine measures in the introduction of poultry or poultry products to the island. Quarantine regulations govern the importation of materials from Macquarie Island to the Australian mainland and vice versa. Strict regulations govern the use of poultry products on the island. It is, however, probably impossible to halt the natural spread of viruses such as Newcastle Disease virus to the island no matter how strict any quarantine measures. Several species of seabird have been involved in epizootics of Newcastle Disease in domestic birds. The virus has been isolated from naturally infected cormorants, from a gannet and has also been reported in nestling cormorants, egrets and herons in the Volga delta.

A possible route by which the virus may have reached Macquarie Island is the tick, *Ixodes uriae*, a parasite of seabirds, widespread in the Arctic, subarctic, Antarctica and subantarctic regions. The tick could be spread in the Antarctic and subantarctic by migratory birds such as skuas which periodically disperse into southern Australian waters.

Doherty *et al.* (1975), outlining the results of virological studies on Macquarie Island, suggested that the Antarctic and Arctic populations of *Ixodes uriae,* and their viruses, may be linked through a chain of other species of *Ixodes* and seabirds on which they are parasitic. They suggested that postulated intermediate links between northern and southern populations of *Ixodes uriae* should be examined for related viruses.

One virus isolated from ticks in a royal penguin colony on Macquarie Island is fatal to fairy penguins which are common along the southern coasts of Australia (Morgan *et al.*, 1978). It seems possible that tick-borne (and perhaps other arthropod-borne) viruses cause deaths amongst bird species on Macquarie Island. The problem is to find dead or moribund birds for virological study, such dead and dying birds being promptly scavenged by skuas and giant petrels.

Morgan *et al.* (1978) also examined southern elephant seals on Macquarie Island for San Miguel Sea Lion virus, implicated in abortions in Californian sea lions *(Zalophus californicus)* and lesions on the flippers of northern fur seals *(Callorhinus ursinus).* This virus is indistinguishable from that causing vesicular exanthema of pigs in the United States, a disease finally eradicated from North American pigs in 1959 but which appears to have found a reservoir in marine mammals. No evidence of the virus was found in southern elephant seals on Macquarie Island. There was no evidence of Influenza A virus or avian influenza virus in the bird population (Morgan, 1988).

Applied microbiology on Macquarie Island

Rabbits on Macquarie Island, a source of food for early sealers, have been considered a major problem since the island became a sanctuary. The role of rabbits in overall island ecology, particularly in vegetation changes and erosion, is still debated (Chapters 5, 6). Investigations into possible reduction of the island's rabbit population have been many.

One attempt at rabbit control involved introduction of a strain of myxoma virus directly into rabbit burrows in an aerosol, and direct inoculation of fifteen rabbits with the virus (Sobey *et al.*, 1973). Reduced rabbit activity followed but no sick rabbits were seen. Poisoning with sodium monofluoroacetate (1080) is effective locally but bait-layers have to carry the poison long distances on foot over difficult terrain. Fumigation of burrows (undoubtedly effective in Europe where rabbits live in large warrens) presents problems on Macquarie Island because of the difficulty of distinguishing rabbit from bird burrows. The situation is made more complicated by the fact the burrow-nesting birds and rabbits often use the same burrow. Burrow fumigation is thus not a viable proposition on the island.

There is no suitable native vector of myxoma virus (so successful in controlling rabbit populations in continental Australia) on Macquarie Island. It was thus considered that biological control of rabbits on the island was not feasible. In tests to determine whether it could act as a possible vector of myxomatosis under subantarctic conditions, *Spilopsyllus cuniculi* (the European rabbit flea) was deliberately introduced to Macquarie Island (Sobey *et al.*, 1973; Table 10.2).

The first batch of laboratory-bred rabbit fleas was released on the island in December 1968. During the summer of 1971–72, 6600 fleas collected

Table 10.2. *Fleas from Macquarie Island and their hosts.*

Flea species	Host species
Parapsyllus magellanicus heardi	Rockhopper penguin, white-headed petrel
Parapsyllus cardinis	Antarctic prion, white-headed petrel
Notiopsylla kerguelensis	Antarctic prion, sooty shearwater
Notiopsylla enciari	White-headed petrel
Neopsyllus fasciatus [a]	Black rat (*Rattus rattus*)
Spilopsyllus cuniculi [b]	Rabbits (*Oryctolagus cuniculus*), cats (*Felis catus*)
Xenopsylla cheopsis [a]	Black rat (*Rattus rattus*)

[a] Probably self-introduced with host.
[b] Deliberately introduced.

from rabbits shot on the island were re-released, 30% of rabbits from the initial flea-release areas being flea-infested (Sobey *et al.*, 1973). It was clear that the European rabbit flea survived and multiplied on the island under subantarctic conditions and that more and more rabbits were becoming flea-infested. A likely vector for myxomatosis had been found.

Between 1968 and 1983, about 240 000 European rabbit fleas were released (Sobey *et al.*, 1973). The flea is now well established in the rabbit population but its spread over the island has been slow. Fleas have been released wherever rabbits are present but their spread is limited to about 200 m around the release site. This limited spread of fleas is obvious in areas of high rabbit density (more than eight rabbits per hectare). In one area studied, rabbits shot 300 m from a point of flea-release were flea-free. The furthest known spread of rabbit fleas from a release point is 1 km in an area of high rabbit density.

The main problem facing island-wide spread of the flea is rabbit behaviour on Macquarie Island. Island rabbits are essentially sedentary, activity throughout the year being confined to a radius of about 200 m from burrows. Rabbits occur in isolated pockets and spread of the rabbit flea depends on flea-infested rabbits (mostly young) moving to adjacent pockets of rabbit infestation. In some years few young survive to disperse as a result of burrow flooding and predation (Copson, Brothers & Skira, 1981).

Rabbit fleas depend on pregnant does for reproduction, another complicating factor. Fleas lay eggs on rabbit nestlings and in rabbit nesting material. New-generation flea pupae emerge as adults when disturbed, such as when female rabbits explore burrows at the start of a breeding season. Flooding of burrows at any stage may lead to egg, larval, pupal and adult flea mortality without necessarily having much effect on rabbits.

Following establishment of rabbit flea populations on the island, myxoma virus was introduced in November 1978 (Brothers, Eberhard, Copson & Skira, 1982) using freeze-dried Lausanne strain virus. Rabbits were inoculated after trapping or were shot with virus-treated air-rifle pellets. The latter method is much less labour-intensive and was adopted as the preferred technique. Holes are drilled in the pellets and plugged with virus-saturated cotton wool. The rifles used are adjusted to a power necessary to penetrate the skin of a rabbit without causing major injury, and shooters aim at the rump to avoid any major damage to the rabbit.

At some sites, inoculation of rabbits with the virus failed to cause the expected epizootics of myxomatosis, probably due to the small number of animals inoculated. The greatest distance over which the virus spread after direct inoculation was 3 km, covered in twelve months. Skira, Brothers & Copson (1983) suggest that the present pattern of rabbit flea distribution means that further artificial spread of the flea and virus is necessary if myxomatosis is to become a more effective control measure on the island. The limited distribution of fleas, over 13 years after their introduction and artificial spread, suggests that other rabbit control measures need to be used in conjunction with fleas and myxoma virus, although the introduction of the virus initially led to a great reduction in rabbit numbers in some areas.

Parasites of elephant seals and birds

Lice and southern elephant seals

Murray (1958) and Murray & Nicholls (1965) studied the sucking louse *Lepidophthirus macrorhini* which infests elephant seals on Macquarie Island. The louse is adapted to life in a habitat (elephant seals) which changes quite abruptly from terrestrial to aquatic and from warm to cold as a seal goes to sea. The habitat additionally sheds both hair and outer layer of the stratum corneum each year.

Lepidophthirus macrorhini is found in greatest numbers on hind flippers of elephant seals, living in burrows in the stratum corneum. Its life cycle consists of five stages: egg, three larval instars and adult. Experiments show that female lice will not lay eggs, nor will eggs hatch, in water. Eggs are attached to hair bases and hatch in five to ten days. Development is rapid and a new generation of adult lice develops within about three weeks. Each female lays an average of six to nine eggs daily. Thus the species can undergo very rapid population growth.

Transfer of lice to new-born seal pups from adults is rapid and the gregarious nature of southern elephant seals when ashore offers ample opportunity for louse transmission from one beast to another. An estimated 87% of southern elephant seals on the island are louse-infested.

There are, however, limits to the growth of louse populations. Females will only lay eggs on seal flippers in air, and reproduction can only occur at temperatures above 25 °C. Temperature measurements of elephant seal hind flippers show that even when ambient temperatures are 1–8 °C, the temperature within folded hind flippers exceeds 25 °C. Lice can thus reproduce quite easily while a seal is ashore.

When the seal goes to sea for months at a time, however, the lice are faced with cold temperatures, immersion in water (because elephant seals do not trap air between their sparse hairs), and the necessity to feed at least once a week to survive. The louse is well-enough adapted to withstand such conditions for some still to be alive when the seal returns to land. These robust individuals establish new louse populations.

When the seal is in water its skin temperature drops to about that of the surrounding water. Lice on the hind flippers, where most vasodilation and heat dissipation occur, have the best chance of survival. At temperatures encountered in subantarctic waters the louse's metabolic rate drops dramatically and its demand for oxygen drops to such an extent that it can obtain sufficient oxygen by cuticular respiration rather than reliance on gas exchange through spiracles. It is possible that the external structure of lice keeps the opening to their burrows open when a seal dives, possibly trapping tiny air bubbles at burrow entrances – artificial gills.

When southern elephant seals moult they shed both hair and the outer layer of the stratum corneum, the skin coming away in great patches (Chapter 9). Any lice or louse eggs on the skin surface are probably lost in the process. Nymphs and adults, in burrows in the stratum corneum, may survive the moult because only the top of their burrows is lost and they can re-establish a population after the moult has finished.

With increasing age, moulting behaviour of southern elephant seals changes (Chapter 9). Pups and young seals moult on the beaches. Mature cows moult in wallows where the louse cannot breed. After the seal moulting period there are fewer lice on adult cows to withstand the next period at sea and repopulate the flippers when they once return to land. The overall louse population, however, is maintained in younger animals.

Ticks, fleas and penguins

Murray (1964), summarising the ecology of ectoparasites of seals and penguins, divided them into two general groups:

(a) those which live permanently on the host (lice, mange mites, feather mites); and

(b) those which visit the host only to feed (fleas and ticks).

The probable importance in spread of viruses by the tick *Ixodes uriae* has been discussed earlier. The only other tick known from Macquarie Island is *Ixodes pterodromae,* parasitic on king penguins, southern skuas and Antarctic prions (Watson, 1967).

Murray & Vestjens (1967) studied the distribution of *Ixodes uriae* and *Parapsyllus magellanicus heardi* on Macquarie Island. The principal hosts of both parasites are penguins. Four species of flea ectoparasites of birds (Table 10.2) are also present. *Ixodes uriae* has a large range of bird host species; *I. pterodromae* has been found only in burrows of Antarctic prions. Flea species which parasitise birds occur in petrel burrows. Only *Parapsyllus magellanicus heardi* is found on penguins. *Ixodes uriae* and *Parapsyllus magellanicus heardi* are the only tick and flea common to penguins on Macquarie Island.

Ixodes uriae is a three-host tick, larvae and nymphs falling to the ground after feeding from the host to moult before the adult tick emerges to feed. Murray & Vestjens (1967) found that the life cycle of *I. uriae* can be completed on Macquarie Island in 270–350 days.

The tick is widespread on Macquarie Island, mostly associated with royal penguin colonies, but is found in localities ranging from the base of tussocks on the edge of royal penguin colonies to nests of rockhopper penguins and rims of cormorant nests. It also occurs in nests of wandering albatrosses, burrows of white-headed petrels on tussock-covered slopes and in burrows of dove prions in feldmark.

In royal penguin colonies the tick occurs in tussocks in and around the colonies, even occurring at depths of 7.5 cm in accumulated layers of feathers and mud blown onto rock faces in colonies. No ticks occur on floors of colonies except where the ground is well-drained and drier. All stages of the tick life cycle are present all year.

Murray & Vestjens (1967) found the heaviest infestations of *Ixodes uriae* on royal penguins, moderate infestations on rockhopper penguins and occasional infestations on king penguins. Gentoo penguins appear to be tick-free. Chicks are more heavily infested than adults. Numbers of ticks on royal penguins vary with the moulting stage. Ticks are absent on birds returning to the island after long periods at sea. Tick infestation is heaviest during the later stages of royal penguin moulting but birds which have finished moulting lack ticks.

Murray & Vestjens (1967) carried out experiments to determine survival rates of various stages of the life cycle of *Ixodes uriae*. They placed larvae and adult males and females in test tubes half-filled with contents of southern elephant seal wallows – a mixture of elephant seal faeces, bacteria, water, and peat. Here, larval and adult ticks survive immersion for several months. Eggs submerged in the same medium failed to hatch. Other experiments showed that female *Ixodes uriae* will not deposit eggs when exposed to wet conditions (water 1–1.5 mm deep) but will do so in moist, non-submerged conditions.

Ixodes uriae succeeds best on ground which is relatively dry. Sufficient offspring are then produced to survive until the host species returns to the island. Close proximity of host species to tick-infested areas is important for survival of the parasite as adult ticks must have a blood meal before they can breed. Thus breeding sites, behaviour and moulting patterns of penguin species parasitised determine distribution and abundance of *Ixodes uriae* on Macquarie Island.

Fleas

Two flea species are self-introduced to the island, both of which parasitise rats. One other has been deliberately introduced, along with myxoma virus. The remaining flea species present are indigenous (Table 10.2).

Parapsyllus magellanicus heardi, which breeds mainly in rockhopper penguin colonies, is the only indigenous flea species which has been studied in detail (Murray & Vestjens, 1967), and rockhopper penguins are the only penguins on the island infested with fleas.

Fleas are common on the brood patches of rockhopper penguins. Sampling of their nesting material, at monthly intervals from June to December, showed that both larval and adult fleas are present throughout the period. Populations of adult fleas and flea eggs increase during the rockhopper penguin nesting period beginning in November. Flea eggs first appear in December. A rockhopper penguin nest sampled in November yielded 100 adult fleas and 50 larvae per 15 cm^3 of nesting material. Collections of *Parapsyllus magellanicus heardi* show that it is only common in rockhopper penguin nests in caves or nests sheltered by rocks. The reason becomes apparent when experiments carried out by Murray & Vestjens (1967) are considered.

Larvae collected from a rockhopper penguin nest in midwinter and placed in petri dishes with nesting material survived until December. Other collections of larvae and nesting material, made in July, were kept at 1–7 °C, 3–9 °C and 3–12 °C. Active larvae were present in all samples in December. Larvae placed in dishes of water and kept at 1–7 °C all died within five days. Adult fleas placed in water at these temperatures died rapidly. It is clear that the flea cannot withstand wet conditions, hence its abundance in sheltered rockhopper penguin breeding sites and absence from water-sodden colonies of other penguin species on the island.

The host range of *Parapsyllus magellanicus heardi* on Macquarie Island is limited and does not reflect the flea's host range in the subantarctic as a whole (see Table 10.3).

Other parasites

Rabbit fleas and myxoma virus are parasites new to rabbits on Macquarie Island. The rabbits first introduced to the island already had parasites in and on them when they arrived. Bull (1960) studied parasites of rabbits on a number of subantarctic islands, based on very small sample numbers. The meaning of variations in distribution of various rabbit parasites between subantarctic islands is still obscure. Only nine rabbits were sampled from Macquarie Island, the livers, gut, fur and skins of the specimens being examined.

Oocysts of *Eimeria stiedae* (a coccidian protozoan) were found in the gall bladders of Macquarie Island rabbits and oocysts of a further four *Eimeria* species, identified only tentatively, were found in rectal faeces. Two species of parasitic nematode were present in gut samples. Skin and fur samples yielded three species of parasitic mites and a sucking louse. Other arthropods were present on the skins, some too damaged to be identifiable. One, a mite common in stored produce, had probably invaded the specimen's fur after the animal's death.

Bull pointed out that all parasites found on rabbits in the subantarctic are one-host parasites with only a brief non-parasitic stage in their life cycle. Mites and lice recorded on rabbits are parasitic throughout their lives while non-parasitic stages of coccidian protozoa and nematodes occur in rabbit faeces and become infective within a few days.

Australian and New Zealand rabbits have an impoverished parasite fauna when compared with rabbits in Europe; an impoverishment further accentuated on the subantarctic islands. The absence of cestodes and liver flukes in rabbits on subantarctic islands is explained by the absence from the islands of secondary hosts necessary for completion of the parasite life cycle.

Table 10.3. *Host range of* Parapsyllus magellanicus heardi *in the sub-antarctic.*

Island(s)	Host(s)
Macquarie	Rockhopper penguin, white-headed petrel
Marion	Macaroni penguin
Kerguelen	Rockhopper penguin, light-mantled albatross, giant petrel, great-winged petrel, blue petrel, Kerguelen cormorant, southern skua
Heard	Light-mantled albatross, Antarctic prion, southern skua, Cape petrel

Data from Murray & Vestjens (1967).

Mawson (1953) described parasitic nematodes collected by expeditions to Heard and Macquarie Islands. She recorded two species from leopard seals on Macquarie Island, two species from southern elephant seals and one from a New Zealand fur seal. Birds from Macquarie Island harboured three nematode species and fish a further three species. Nematodes in fish from Macquarie Island were encapsulated within the body cavity.

Free-living arthropods

Free-living terrestrial arthropods of Macquarie Island have been little studied except from the viewpoint of taxonomic description and listing of collembolan, insect and arachnid species present (Watson, 1967; Gressitt *et al.*, 1962; Greenslade, 1987a). Ecological studies have essentially been confined to the parasitic species dealt with above and an intensive study of the copepod *Pseudoboeckella brevicaudata* (Chapter 7).

Kelp flies

An outstanding feature of the shore environment of Macquarie Island is the huge piles of rotting kelp which litter the beaches after storms. Two kelp fly species are important in this zone: *Coelopa curvipes* and *Coelopa debilis* (Watson, 1967). During high winds and low ground temperatures, adult *Coelopa curvipes* gather in thick masses within piles of kelp or hide under pebbles. In severe storms the flies shelter in tussocks. Larvae of both species are found on and within masses of decaying kelp but larvae of *Coelopa curvipes* also occur in huge numbers in decaying carcasses of elephant seals and in the sand beneath them. *Coelopa curvipes* larvae are always outnumbered by those of *Coelopa debilis* in decaying kelp, the reverse situation obtaining in rotting seals. Larvae of the two species obviously have an important role in breaking down both beach-washed kelp and dead seals. The mixture of rotting kelp (much favoured by hauled-out elephant seals) with a mulch of seal excrement and masses of microbial slime leads to development of huge numbers of larvae which then pupate on the sand beneath the kelp or on kelp fronds which have been only partly decomposed. The larvae are probably more important in earlier stages of kelp decomposition than in later stages (Bunt, 1955).

Appendix 7 contains a checklist, prepared by P. Greenslade, of arthropod species known from Macquarie Island.

11

The nearshore environment

Intertidal ecology

The intertidal and upper subtidal regions of the rocky shores of Macquarie Island have been the subject of a series of studies on the ecology of their fauna and flora (Kenny & Haysom, 1962; Bennett, 1971; Simpson, 1972, 1976a, b, 1977, 1982a, b, 1984, 1985; Simpson & Harrington, 1985).

Physical factors, such as aspect, temperature, desiccation and salinity, as well as biotic factors, including food availability, predation, mode of reproduction, and degree of shelter provided by marine algae, have a marked effect on distribution and abundance of such littoral molluscs as *Cantharidus coruscans*, *Laevilittorina caliginosa*, *Kerguelenella lateralis*, *Plaxiphora aurata*, *Hemiarthrum setulosum* and *Patinigera macquariensis* (Simpson, 1976a, b, 1977, 1985).

Habitats and zonal sequences of species are generally related to their tolerances of the factors outlined above. One exception is that heat resistance in chitons, when compared with that of gastropods, is greater than their range of distribution would suggest. Biotic factors influencing distribution and abundance of species may reflect the comparative stability of the physical environment. The chitons *Plaxiphora aurata* and *Hemiarthrum setulosum* show variable species morphology, mobility, food preference, degree of shelter sought amongst the large kelp *Durvillaea antarctica*, predation by birds (Dominican gulls) and isopods, and a brooding mode of reproduction.

Molluscs in the littoral zone face a number of hazards on subantarctic shores: dislodgement from the substrate during heavy seas and subsequent deposition at higher levels, and burial under dense layers of stranded kelp resulting in death of underlying animals. In summer, moulting elephant seals form groups in suitable areas in the coastal fringe, particularly on stranded kelp. The overlay of seals, their excrement and rotting kelp quickly kills any rock pool organisms (Simpson, 1976a).

Kenny & Haysom (1962) recognised, on the rocky shores of the northern part of the island, six zones extending from high-water spring tide to below low-water spring tide levels: a lichen zone, a *Porphyra* zone, a 'bare' zone dominated by limpets, an upper red algal zone, a *Durvillaea* zone, and a lower red algal zone.

While Kenny & Haysom (1962) related zonation patterns solely to the local habitat at Macquarie Island, Simpson (1976b) attempted to relate local zonations to a universal zonation scheme proposed by Lewis (1961, 1964). The extent of any zone is related to coastal site exposure, slope profile and substrate stability (Figure 11.1). The scheme is, however, based on a simple slope model and is not valid where topography varies significantly from a simple slope. As Simpson (1976b) stated:

> This exemplified the complexity of factors modifying the zonation pattern and emphasized that the total set of conditions controlling zonation is best revealed by the organisms themselves.

With the exception of the 'bare' zone, dominated by the limpet *Kerguelenella lateralis,* the zones are dominated by algae. Species lists of the flora and fauna of all zones are given by Kenny & Haysom (1962) and Simpson (1976b). Based on their descriptions, the six zones are described below.

1. Lichen zone

This zone (Figure 6.1) is dominated by lichens, principally species of *Verrucaria, Mastodia, Lecanora, Xanthoria* and *Buellia.* The mosses *Muelleriella crassifolia* and *Pottia heimii* are common and, together with the vascular plants *Colobanthus muscoides, Cotula plumosa* and *Puccinellia macquariensis* (Figure 6.2), provide a refuge for invertebrates, particularly mites and collembolans. Insects, including flies, a midge, beetles and a wasp, occur in this zone. A copepod, *Tigriopus angulatus,* and amphipods have been recorded along with a number of algae, including *Hildenbrandia lecannellieri, Prasiola crispa* and *Enteromorpha* spp.

The upper boundary of the lichen zone grades rather confusedly into the terrestrial maritime communities of rocky shores (Chapter 6). At its lower limit there is generally a sharp transition between lichen-dominated and *Porphyra*-dominated vegetation.

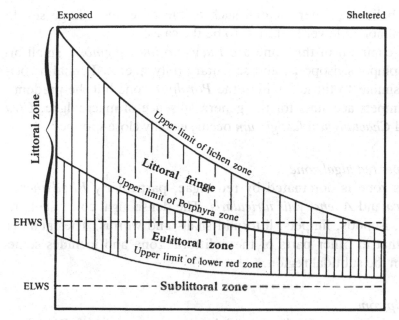

Figure 11.1. Zonation of rocky shores on Macquarie Island, related to the universal zonation scheme. ELWS = extreme low water spring tide, EHWS = extreme high water spring tide. Redrawn from Simpson (1976b).

2. *Porphyra* zone

Porphyra columbina dominates this zone, forming dense swards, the upper limit of which is determined by high-water spring tides. This zone is frequently wet by wave splash. Other algae such as species of *Enteromorpha, Cladophora, Ulva* and *Prasiola crispa* occur here. Mites, oligochaetes, sedentary polychaetes, *Laevilittorina caliginosa,* small bivalves, amphipods, *Exosphaeroma* (an isopod) and the limpet *Kerguelenella lateralis* are found sheltering amongst the algae.

3. 'Bare' zone

Kenny & Haysom (1962) described considerable variation in both horizontal and vertical extent of the zone dominated by the limpet *Kerguelenella lateralis*. Grazing of epilithic algae by *Kerguelenella* may help maintain an apparently 'bare' rock surface. The limpet favours dissected or creviced rock surfaces. Kenny & Haysom (1962) reported limpet densities of up to 1500 per square metre. They believed the scouring action of fronds of *Durvillaea antarctica* to be the main cause of the relatively barren nature of rocks in this zone, but Simpson (1976b) disagreed, believing other factors, such as the absence of large sessile animals common at lower latitudes, to be the cause.

Animals common in this zone are *Laevilittorina caliginosa,* small bivalves, amphipods, isopods, and sedentary polychaetes. Faunal associations are similar to those found in the *Porphyra* zone but the predominance of limpets accounts for the general absence of macroalgae. *Ulva* species and *Chaetangium fastigiatum* occur, usually close-cropped.

4. Upper red algal zone

This zone is dominated by red algae, particularly *Rhodymenia subantarctica* and *Adenocystis utricularis.* The dense algal cover shelters amphipods, isopods, limpets, littorinid snails and worms. The fauna is similar to that in other parts of the eulittoral zone and includes some species from the subtidal region.

5. Kelp zone

The kelp zone is dominated by massive plants of *Durvillaea antarctica* whose fronds, by their movement with waves and tides, affect zones above and below them. Areas between the holdfasts of this alga are covered by coralline red algae. Eroded holdfasts of *Durvillaea* support a variety of animal species (Figure 11.2).

Few animals occur in the coralline red algal zone in areas where kelp has been removed naturally. Simpson (1976b) experimentally removed holdfasts of *Durvillaea* and showed that most of the animals normally present died or migrated away from the cleared area. Numbers of the limpet *Patinigera macquariensis* increased.

Areas around the holdfasts of *Durvillaea* form a habitat sheltered from all but major wave action, high light intensity and solar radiation, wind and heavy rain. The zone of giant kelp plays an important role in modifying the effects of wave action around the coast, absorbing much of the wave energy and protecting the land from the full force of wave action.

Two distinct forms of *Durvillaea antarctica* are recognisable. In plants of the intertidal region the lamina is spongy with an air-filled medulla. The subtidal form, found at depths of 15 m or more, has compressed laminae with small air spaces (Klemm & Hallam, 1988).

6. Lower red algal zone

Below the kelp zone is a zone dominated by coralline and other red algae with an extensive and varied fauna.

Figure 11.2. Animals such as chitons and limpets are found amongst the stipes of *Durvillaea antarctica*. The rock surface is encrusted with coralline red algae below the *Durvillaea antarctica* stipes. Bauer Bay.

Sublittoral organisms

During the southern summer of 1977–78 a survey of the inshore marine flora and fauna was made by a team from the National Museum, Sydney, supported by SCUBA diving (Lowry, Horning, Poore & Ricker, 1978). This was the first systematic collection of subtidal organisms made on Macquarie Island. Specimens collected are still under study. Many collections of species previously unknown from Macquarie Island were made, including members of the Trematoda, Opisthobranchia, Polychaeta, Amphipoda, Isopoda, Pycnogonida, Echinodermata and Ascidiacea. The collections, when finally worked up, will greatly help our understanding of biogeographic relationships of the island's marine invertebrates.

Marine algae

The marine algal flora of Macquarie Island was initially studied in the intertidal region, supplemented by collections from drift and a few subtidal collections (Zinova, 1958, 1963; Kenny & Haysom, 1962). Some 59 species of red, brown and green algae were reported from the littoral zone. Subtidal studies carried out during the summer of 1977–78 raised the number of marine algae recorded from Macquarie Island to 103 species (Ricker, 1987).

Brown algae, although lower in diversity than red algae, dominate most of the algal communities. The three major brown algae *(Durvillaea antarctica, Macrocystis pyrifera, Desmarestia* spp.) supply shade and shelter for low-light-tolerant red algae, many invertebrates and some fish species, and serve as a substrate for numerous epiphytic algal species and invertebrates. Green algae are less diverse than the brown algae but some (such as *Codium subantarcticum)* may dominate subtidal communities in sheltered areas. Green algae (such as species of *Enteromorpha, Ulva, Rhizoclonium ambiguum* and *Prasiola crispa)* dominate some intertidal communities, particularly in the upper sublittoral zone where the supply of freshwater appears to correlate with the dominance of green algae. The freshwater inhibits or prevents growth of less tolerant red and brown algae.

Changes in the type of algal dominance in intertidal pools on a rock platform on the east coast south of Green Gorge show the effects of salinity on algal community structure (Table 11.1).

The oceans around Macquarie Island

The Southern Ocean supports a highly productive ecosystem in which krill *(Euphausia* spp.) occupy a key role. Many of the major carnivores in the oceanic food chain – whales, seals, penguins and other seabirds, cephalopods, fish and some zooplankton – rely directly or indirectly on various stages of the krill life cycle as a source of food. Estimates of the biomass of krill in the Southern Ocean vary widely. The United Nations Food and Agriculture Organization estimates range from 125 million to 750 million tonnes of krill (Fischer & Hureau, 1985). It has been suggested that the drastic reduction in populations of baleen whales following their exploitation for blubber and whalebone released about

Table 11.1. *Changes in algal dominance with salinity in rock pools on rock platform near Green Gorge.*

Pool	Salinity (ppt)[a]	Vegetation
High intertidal pools		
Very small volume 15 cm deep	18	Filamentous greens *Enteromorpha* spp.
Small volume 45 cm deep	19	Filamentous greens *Enteromorpha* spp.
Medium volume 45 cm deep	20	Filamentous greens *Enteromorpha* spp.
Large volume 75 cm deep	22	Dense mat of filamentous greens
Large volume 90 cm deep	24	Filamentous greens Encrusting corallines *Hildenbrandia lecannellieri* (red) *Adenocystis utricularis* (brown) (few)
Very large volume	32	Encrusting corallines *Adenocystis utricularis* *Hildenbrandia lecannellieri* Foliose red algae
Low intertidal pools		
Small volume 45 cm deep	28	Encrusting corallines *Hildenbrandia lecannellieri* *Adenocystis utricularis*
Large volume	33	Encrusting corallines *Hildenbrandia lecannellieri* Foliose reds
Very large volume receiving regular wave wash	33.8	Encrusting corallines *Hildenbrandia lecannellieri* *Adenocystis utricularis* Filamentous and foliose reds

[a] Open sea salinity = 33.8 ppt.

Modified from Lowry *et al.* (1978).

100–150 million tonnes of krill which could be harvested annually by humans without depleting stocks. A decline of one predator, however, usually results in faster growth rates of other predators, their earlier maturity, increased fertility and survival, and larger populations. An example is the rapid increase in population size of fur seals on South

Table 11.2. *Sea temperatures (°C), Macquarie Island.*

Period	Monthly mean												Annual mean
	Apr.	May	Jun.	Jul.	Aug.	Sep.	Oct.	Nov.	Dec.	Jan.	Feb.	Mar.	
1912–14[a]	4.72	3.83	3.44	3.28	3.44	3.67	3.89	4.83	4.50	6.33	5.50	5.06	4.50
1951–54[a]	5.78	4.72	3.78	3.61	3.50	3.78	3.83	5.44	6.72	7.28	7.22	7.22	5.22
1968–69[b]	5.61	5.72	5.28	4.28	4.67	4.50	4.33	4.56	5.72	6.78	6.67	6.28	5.39

[a] Converted from °F after Loewe (1957). Recording time = 0900 hours.
[b] Recording time = 0900–1100 hours.

From Simpson (1976b).

Georgia. Similar increases in population size may have occurred in some seabird and whale species and possibly squid and fish populations in the Antarctic and subantarctic.

Krill and some other crustaceans, fish and cephalopods are actively harvested in a number of regions in the Southern Ocean. The greatest exploitation is concentrated in the Weddell and Scotia Seas, where stocks of krill are highest. Fish, cephalopods and crabs are harvested around Iles Amsterdam, Iles Crozet and Iles Kerguelen. Over-exploitation near Iles Kerguelen is a real possibility. There is evidence to suggest the presence of an oceanic fishery in waters around Macquarie Island. There is an upwelling of nutrient-rich Antarctic waters close to the south-eastern end of Macquarie Island which is probably reflected in the abundance of breeding royal and king penguins on this part of the island.

Over-exploitation of marine living resources for human needs can result in loss of species and disruption of the natural balance of the ecosystem. Under the auspices of the Commission for the Conservation of Antarctic Marine Living Resources (CCAMLR), a multinational research programme on the Antarctic marine ecosystem and its living resources was established – the Biological Investigation of Marine Antarctic Systems and Stocks (BIOMASS) programme. Antarctic krill, *Euphausia superba,* was the central organism for this study. The first BIOMASS experiment (FIBEX) concentrated on krill abundance and distribution while the second BIOMASS (SIBEX) was designed to study seasonal distribution patterns of krill and other organisms, abundance and production related to the physical environment, food availability, competitors and predators.

It has become apparent that early estimates of krill abundance were inaccurate. Fischer & Hureau (1985) remarked that our present knowledge of krill is inadequate for resource management purposes and the great variation in early estimates of abundance and production make it impossible to determine changes in abundance of krill which may have already occurred (due to exploitation of whales, or krill itself) or changes which may yet occur.

In addition to krill, fish populations in the Southern Ocean have recently been subject to commercial exploitation. About 270 species are known but only about 25 are of commercial interest (Fischer & Hureau, 1985). Juvenile fish, if not adults, are important in the diets of seabirds and seals. Estimates of recent harvests in the Southern Ocean are approximately 100 000 tonnes per year around South Georgia, the South Orkney and South Shetland Islands and 20 000 tonnes around Iles Kerguelen. As much as 100 000–230 000 tonnes were harvested annually at the onset of the

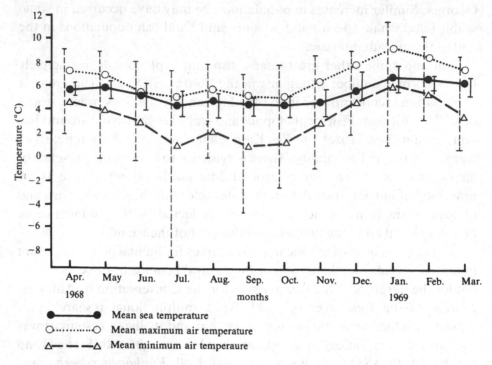

Figure 11.3. Mean sea and air temperatures at Macquarie Island. Solid vertical bars give range of air temperatures, dotted vertical bars give range of sea temperatures. Monthly April 1968 to March 1969. Redrawn from Simpson (1976b).

fishery around Iles Kerguelen. Maximum sustainable yield estimates have been made at around 20 000 tonnes annually.

Cephalopods, particularly squid, are important in the diet of the larger seabirds (albatrosses, penguins), seals and some whale species. Little is known about their abundance or biology in the Southern Ocean.

Exploitation of elephant seals ceased in most subantarctic localities early in the twentieth century. On South Georgia controlled exploitation continued between 1910 and 1964, with only adult males being taken. The population increased considerably (Table 9.3). Since the 1950s there has been a general decline in numbers of elephant seals at least on Marion, Heard and Macquarie Islands. The reasons are still unclear. On Iles Kerguelen the population fluctuates but remains, on the whole, stable.

Fur seals were hunted almost to extinction in the nineteenth century. Their numbers are now increasing throughout the subantarctic region. A dramatic increase in numbers has occurred on South Georgia. Reasons for this are, again, not understood. Competition for ecosystem resources, at least with seabirds, may be expected. On South Georgia fur seals have expanded their haul-out habitat from rocky beaches and have destroyed or damaged significant areas of surrounding vegetation.

The indigenous vertebrate fauna of Macquarie Island, seals and seabirds, depends on the productivity of the surrounding ocean. The terrestrial flora receives a significant input of wind-borne sea-salt (see Chapter 7) as well as nutrient input from breeding birds and, on the coastal fringe, seals.

The vast expanses of ocean surrounding the island affect, directly or indirectly, many aspects of the environment. Mean seawater temperatures fluctuate between 7.28 °C in summer and 3.28 °C in winter (Table 11.2; Figure 11.3).

Seas surrounding Macquarie Island are deep (Figure 4.3). Soundings of 1548 fathoms (2830 m) have been made as close as 5 nautical miles (9.3 km) from the eastern shore (Mawson, 1943). The nearest islands, Auckland and Campbell Islands, are separated from the Macquarie Ridge by depths of more than 2000 fathoms (3600 m) (Chapter 4).

Whales

Six species of whale have been recorded from waters near Macquarie Island (Anon, 1987). Cuvier's beaked whale *(Ziphius cavirostris)* has been found as a beach-washed skull. The southern bottle-nose whale *(Hyperoodon planifrons)* is known from one stranded specimen. Southern right whales *(Balaena glacialis)* and sperm whales *(Physeter macrocephalus)* are occasionally sighted from the shores of the island. Longfin pilot whales *(Globicephala melaena)* and killer whales *(Orcinus orca)* are known from both strandings and sightings. Killer whales are the most frequent cetacean visitors to island waters, at least during summer when small pods are often sighted, close inshore, probably feeding on penguins and immature elephant seals. The impact of whales on potential food resources for other ocean-feeding inhabitants of Macquarie Island is probably minimal.

Fish

Until recently only seven benthic and two or three pelagic (open-ocean dwelling) fish species had been reported from Macquarie Island. Many of the fish species recorded were initially obtained from beach-washed specimens, penguin stomach samples, caught on lines from the shore or in shallow-water fish traps. Waite (1916) published the first account of fish caught near or from Macquarie Island. Some species obtained from open-ocean collections were included.

Nearshore surveys including trawls conducted in 1986 increased the number recorded to 12 benthic and 21 pelagic species (Williams, 1988) (Appendix 9). The surveys are as yet incomplete. The number of benthic species is low compared with records from Iles Kerguelen (36 benthic species) and South Georgia (28 benthic species). However, only a small area of the ocean surrounding Macquarie Island is shallower than 1000 m.

The paucity of fish species around Macquarie Island can be explained by the expanse of deep ocean which separates Macquarie Island from any other land mass, limiting the chance for shallow-water bottom-dwelling fish to reach the island. Further, the easterly current drift limits the probability of the pelagic young stages of fish from the Australasian region reaching the island.

Biogeographically, the Macquarie Island fish fauna is allied with that of the Kerguelen region. Poor collections make any biogeographic conclusions difficult. Of the twelve benthic species recorded, five are common to the Australasian region; seven of the species are common to the Kerguelen region; six are known from the Scotia region.

The fish fauna is overwhelmingly subantarctic in affinity, but there are some basically temperate and some basically Antarctic species within the fauna. This may be expected from the position of the island relative to the Antarctic convergence (Chapter 3). Members of the AAE reported a beach-washed sleeper shark *(Somniosus microcephalus)* (Waite 1916). The shark has a world-wide distribution.

Muraenolepis marmoratus has a wide Antarctic distribution. *Zanchlorynchus spinifer, Dissostichus eleginoides, Notothenia* spp., *Paranotothenia* spp., and *Harpagifer georgianus* are common in the subantarctic region. *Paranotothenia magellanica* is readily caught close to the shore while *Notothenia squamifrons,* the most common species in recent netting trawls (Williams, pers. comm.), is commercially exploited around Iles Kerguelen. *Notothenia rossii* has been heavily harvested in waters around South Georgia. *Centriscops humerosus, Neophrynichthys magnicirrus* and *Ebinania macquariensis* are Australasian in distribution. *Paranotothenia magellanica,* principally subantarctic in distribution, is also found in New Zealand waters.

Little is known of the biology of the fish of the region. Both *Harpagifer georgianus* and *Paranotothenia magellanica* are commonly encountered in the kelp zone fringing the island. Nototheniid fish at Macquarie Island, as elsewhere, appear to feed mainly on benthic and some pelagic crustaceans such as amphipods, isopods and euphausiids.

Fish, particularly *Harpagifer georgianus* and small *Paranotothenia magellanica* form almost the entire diet of the Macquarie Island cormorant (Brothers, 1985; Chapter 8). Fish are also an important component of the diet of gentoo and royal penguins. Rand (1956) and Condy (1978) point out that fish are important in the diet of the subantarctic fur seal, *Arctocephalus tropicalis*.

Scavengers such as skuas, dominican gulls, giant petrels and wekas take beach-washed fish. Merilees (1984b) reported a mass stranding of lantern fish, *Electrona subaspera* and *Gymnoscopelus braueri,* both usually associated with deep water. Dominican gulls were seen feeding on the stranded fish. Merilees suggested that the mass stranding may have been the result of a sudden lowering of seawater temperature by about one degree as a transient body of colder Antarctic water surrounded the island.

Cephalopods

Cephalopods form a significant part of the diet of whales, seals, albatrosses and some penguins (see Chapters 8 and 9). Little is known about the species which inhabit waters around Macquarie Island. Occasional mass strandings have been reported and isolated specimens are found beach-washed. O'Sullivan, Johnstone, Kerry & Imber (1983) reported a mass stranding of *Martialia hyadesi* at the northern end of the island in 1971, suggesting that a sudden change in wind direction, coupled with a rising tide, may have been contributing factors to the stranding. Fischer & Hureau (1985) indicate that other species, *Todarodes filippoarae* and *Moroteuthis ingens,* may inhabit waters around Macquarie Island.

Crustacea

The subantarctic stone crab *(Lithodes murrayi)* is known from the Prince Edward Islands, Iles Crozet, the Kara Dag seamount east of Iles Crozet, Macquarie Island, southern Chile, south-west Africa (as far north as 22°S), Natal (off Durban) and from the Foveaux Strait, New Zealand. It is common around Iles Crozet and in New Zealand waters.

Around Iles Crozet, *Lithodes murrayi* has been recorded, according to season, from the shoreline to a depth of 1015 m. The animals migrate seasonally to shallow waters for breeding and then return to deeper waters. The crab is an opportunistic feeder with a diverse diet, including algae, invertebrates, carrion, and penguin feathers. It has been caught commercially off south-west Africa and around Iles Crozet and is being taken in increasing numbers in New Zealand. The industry at Iles Crozet was abandoned as uneconomic after only a few years due to depletion of the stock. O'Sullivan (1983a) suggested that the crab population at Macquarie Island is not suitable for exploitation.

Kirkwood (1982), in his handbook of Euphausiid species of the Southern Ocean, included three species, all widespread, both north and south of the Antarctic convergence, likely to occur in waters around Macquarie Island. *Euphausia triacantha* is circumpolar between latitudes 50°S and 60°S, but more common just south of the Antarctic convergence. It occupies depths of 250–750 m by day, rising to above 250 m by night, but does not swarm. *Euphausia valentini* is also circumpolar, on both sides of the Antarctic convergence, although chiefly north of it. It forms large swarms at 100–250 m depth by day, rising to near the surface at night, and is an important food of seabirds. Royal penguins rely heavily on this species for food (see Chapter 8). *Thysanoessa vicina* is rare and little known. It occurs on both sides of the Antarctic convergence, often in the upper 250 m, but has been found as deep as 1000 m. Tattersall (1918) reported immature specimens of *Thysanoessa gregaria* caught at 26 m in a tow net. Sheard (1953), discussing the general distribution of Euphausiacea, stated that *T. gregaria* occurred in warmer waters, between 2°S and 43°S.

Many other groups of marine organisms with both pelagic and benthic stages have been reported from inshore waters at Macquarie Island or the surrounding ocean. Much of our knowledge of these is based on two collections, those of the AAE, 1911–14, and BANZARE, 1929–31. Represented in the collections are: copepods, cladocerans, ostracods, pycnogonids, sponges, jellyfish, flatworms, nemertean worms, nematodes, parasitic Acanthocephala, polychaetes, oligochaetes, amphipods, Endoprocta, chaetognaths, tunicates, echinoderms, and molluscs.

Biogeographic discussions of the marine and littoral faunas are tenuous due to as yet insufficient collections. Based on studies of molluscs, Tomlin (1948) concluded that there was a stronger affinity with Antarctic regions than with New Zealand. Powell (1957, 1971) revised Tomlin's work and considered that there was a stronger alliance with the Kerguelen region to the west. Dell (1964) regarded the Macquarie Island marine fauna as strongly mixed, having close links, particularly among the echinoderms, with New Zealand, as well as other strong affinities with the Kerguelen region and Antarctica. Dawson (1988), like Dell, believes the Macquarie Island marine fauna to be distinctive. He feels designation of Macquarie Island as a separate subantarctic province may be warranted.

12

Human effects: from mismanagement to management strategies

Recorded exploration and contact with high southern latitudes began in the seventeenth and eighteenth centuries with La Roche's, Bouvet's and Kerguelen's voyages and continued with James Cook's extraordinary voyage of 1774–75 during which he crossed the Antarctic Circle three times. Thereafter followed a period of further exploration and intense exploitation based on seals, whales and penguins; a period during which the islands of the subantarctic region were progressively discovered and exploited by humans and colonised by alien plants and animals. During this early exploitation no attempt was made to conserve or manage wildlife stocks for long-term harvesting. Consequently, fur seals were almost exterminated in southern regions and southern elephant seal populations were depleted on some islands.

Sealers accidentally introduced rats and mice to many subantarctic islands. A few, such as Heard and McDonald Islands and Prince Edward Island, were never colonised by these rodents. Sealers deliberately introduced cats to many islands as pets or to control rodents.

Cats have had significant effects on native faunas of all islands to which they were introduced. On Macquarie Island, at least two bird species were exterminated (see Chapter 8) and numbers of some burrow-nesting petrels were considerably reduced by cats or other introduced predators, principally wekas. Some petrel species may now no longer breed on the island (Table 8.4).

Other species introduced to southern regions, such as European rabbits, cattle, sheep, mouflon and reindeer have considerably altered vegetation structure of at least parts of the islands to which they were introduced. Grazing and trampling, particularly by hoofed species, have led to significant changes.

Holdgate & Wace (1961) and Holdgate (1966) provide detailed discussions of human influences on ecosystems of the southern islands.

Johnstone (1985) discussed threats to native breeding populations of seabirds posed by introduced animal species.

The nature of island ecosystems has been well documented (Holdgate & Wace, 1961; Wace, 1982; Clark & Dingwall, 1985). They are quite distinctive and differ markedly from those of continental land masses.

> Islands reflect the overwhelming influence of their oceanic surroundings (especially in their climatic regime) and are characterised by limitations of space, restricted habitats, impoverished faunas and floras compared to continental areas of similar ecological diversity, and a high degree of endemism stemming from their geographical and ecological isolation.   (Clark & Dingwall, 1985)

These distinguishing characteristics of islands are responsible for their intrinsic values as protected areas. Paramount among these values is the uniqueness of their floras and faunas due to the presence of endemic, relict and/or specialised species. Their isolation means that islands are ideally suited as refugia for threatened plants and animals (provided suitable ecological niches are available) and as reservoirs preserving genetic resources.

Because oceanic islands are unique, there are considerable difficulties in managing them to preserve their uniqueness. Their very isolation was for many millennia their best means of preservation.

Accidental or deliberate introduction of alien animals, particularly mammals, and plants is one of the greatest threats to the stability of an island's ecosystem. Macquarie Island had no native mammalian herbivores or terrestrial carnivores before its discovery. Rabbits and cats, more than other introduced vertebrates, have had significant effects.

> Experience reveals that the natural environments of these southern oceanic islands are readily disturbed and destroyed but virtually impossible to rehabilitate or replace. Protected area managers have an awesome responsibility to secure island protected areas against the deleterious influences of man. In recent years the expansion of commercial interests in fishing, mineral (especially oil) exploration and tourism, and increased scientific activity, are inexorably eroding the isolation of the southern islands and pose problems for their effective management as protected areas.   (Clark & Dingwall, 1985)

Introduced animals and plants on Macquarie Island

Introduced animals
Only three subantarctic islands are free of introduced mammals (Table 12.1). All mammals introduced to subantarctic islands have had

Table 12.1. *Alien mammals on subantarctic islands.*

N = naturalised alien; T = transient alien; – = not recorded.

Island	Cats	Rabbits	Brown rats	Black rats	Mice	Reindeer	Domestic[a]	
Macquarie	N	N	–	N	N	–	T	(1)
Heard	–	–	–	–	–	–	?	
Macdonald	–	–	–	–	–	–	–	
Kerguelen	N	N	?	N	?	?	?	(2)
Marion	N	–	–	–	N	–	T	(3)
P. Edward	–	–	–	–	–	–	–	(3)
S. Georgia	T	T	N	T	N	N[b]	T	(4)

[a] Includes sheep, cattle, pigs, goats, dogs, horses, donkeys.
[b] Deliberately introduced to island. Herds subject to continuing study.

(1) Jenkin, Johnstone & Copson (1982). (2) Lesel & Derenne (1975). (3) Smith (1987a). (4) Headland (1984).

substantial effects on the islands they now inhabit. Cats, rabbits, mice and black rats are now naturalised on Macquarie Island.

Cats

Feral cats are a greater threat to survival of small ground-dwelling or burrow-nesting birds than any other introduced mammal on subantarctic islands. At least eight subantarctic islands (some of them members of archipelagos) now have free-ranging cat populations (van Aarde, 1979).

The cat population on Marion Island has been studied intensively for many years. (van Aarde, 1986; van Rensburg, 1986; van Rensburg, Skinner & van Aarde, 1987). It has been estimated that each of the approximately 2000 cats on Marion Island in 1975–76 killed some 213 birds per year, resulting in as many as 455 000 birds being killed annually (van Aarde, 1979, 1980b). Island populations of breeding birds subject to such predation pressure are unlikely to survive in the long term.

Control measures introduced on Marion Island have led to a major decline in cat numbers. In 1949, five cats were introduced to the island to control mice at the scientific station. By 1975, the cat population was estimated at about 2000 (van Aarde, 1979). In 1977, when the population was about 3400 (van Aarde & Skinner, 1982) infectious feline panleuco-paenia virus was introduced as a biological control measure. The cat population fell rapidly to about 600 animals (van Rensburg, Skinner & van

Aarde, 1987). A programme was introduced in the summer of 1986–87 aimed at eradicating all female cats.

Cats are the most significant introduced mammalian predator on Macquarie Island. Jones (1977) estimated their numbers at between 350 and 500. Cats have been implicated in the extinction of two native bird species and kill significant numbers of burrow-nesting petrels (Brothers, 1984). A hunting programme, resulting in the death of hundreds of cats, has apparently not significantly reduced (based on annual numbers of cats shot or gassed) the cat population. Control measures such as those employed on Marion Island may prove far more effective and cost-efficient.

Rabbits

Rabbits were introduced to a number of subantarctic islands as a source of fresh meat (additional to seal and bird meat) for sailors and sealers (Holdgate & Wace, 1961). Their effects on island ecosystems have been considerable. Taylor (1955a) and Copson (1984) discussed effects of rabbit grazing on the vegetation of Macquarie Island. By providing an alternative winter food source for cats, rabbits have been implicated in the extinction of the Macquarie Island parakeet (see p. 135), and at least some of the populations of burrow-nesting birds on the island (Brothers, 1984). Rabbits also provide an additional source of food for skuas, and the distribution of breeding skuas appears to be closely correlated with rabbit distribution (Jones & Skira, 1979).

Rabbit grazing has been blamed for catastrophic erosion in some parts of the island (Costin & Moore, 1960). More recent studies (Selkirk, Costin, Seppelt & Scott, 1983) show that the suggestion of extensive erosion caused by rabbits was based largely on misinterpretation of landscape features and that natural erosive forces were, and still are, responsible for much of the apparent landscape degradation.

In 1968, the European rabbit flea *(Spilopsyllus cuniculi)* was introduced to Macquarie Island (Skira, Brothers & Copson, 1983) as a vector for myxoma virus, successfully introduced in 1978 (Chapter 10). A significant reduction in rabbit numbers has occurred on Macquarie Island. Vegetation recovery, especially of *Pleurophyllum*-dominated herbfield, has been particularly marked in the northern third of the island (Chapter 6). *Uncinia* spp., *Cerastium fontanum*, *Stilbocarpa polaris* and *Poa foliosa* all appear to have become more abundant. The continuing task of reducing rabbit numbers, with the goal of eventual eradication, will become harder as the vegetation, particularly tall tussock grassland, recovers from earlier degradation and rabbit populations become more confined to a number of isolated pockets in areas of thick vegetation and difficult terrain.

Rodents

Black or ship rats *(Rattus rattus),* brown or Norway rats *(Rattus norvegicus*) and mice *(Mus musculus)* reached many subantarctic islands in shipwrecks or cargo and provisions for shore parties. The extent to which rats are predators of seabirds is not clear. Brown rats are more aggressive and larger than black rats. Pye & Bonner (1980) showed that on South Georgia brown rats eat mostly plant matter, invertebrates and carrion. There was also some evidence that they catch birds. There is a negative correlation between distribution of rats and breeding Antarctic pipits *(Anthus antarcticus)* and South Georgian pintails *(Anas georgicus)* on South Georgia. Some burrow-nesting petrel species also seem to be eaten by rats. Black rats have markedly affected distribution and abundance of a number of burrow-nesting birds on Ile Amsterdam and Ile Saint Paul (Segonzac, 1972) Iles Kerguelen (Dorst & Millon, 1964) and on Iles Crozet (Mougin, 1969; Johnstone, 1985).

Black rats and mice are widespread on Macquarie Island, occupying all habitats except feldmark. Diet of both mice and rats on Macquarie Island has been studied (Chapter 9). There is only circumstantial evidence to implicate rats in disturbance to breeding burrow-nesting birds on the island (Brothers, 1984).

Birds

Four bird species alien to the original bird fauna now breed on Macquarie Island (Tables 8.1, 8.2; Appendix 10). Self-introduced bird species have no known effects on the island's ecology. Mallard, originally introduced to Australia and New Zealand, have spread to Macquarie Island and interbreed with native Black duck (Norman, 1987; Norman & Brown, 1987). Whether such interbreeding will have any effects on the island's ecology is unknown, but would seem highly unlikely.

Wekas ('Maori hens') (Figure 2.3) were introduced to Macquarie Island from Stewart Island as a supplementary food source for oil collecting gangs (Table 8.2). Their introduction had marked ecological effects. They became firmly established in breeding populations and spread rapidly over the island. Hamilton (1895) noted that the birds were common. Wekas on the island live mostly in tall tussock grassland on the coastal fringe (Sobey *et al..* 1973). Taylor (1979) and Brothers (1984) thought that wekas had

had a major impact on native bird species, either directly as predators or because they help maintain cat populations by providing an additional winter food source for cats. Brothers & Skira (1984) found major concentrations of wekas along the coastal fringe of the northern half of the island. They estimated the total weka population at up to 500 birds.

Dietary surveys, based on gizzard contents, show that wekas are omnivorous, eating vegetable matter and invertebrates. Kelp fly larvae form a large proportion of their diet. Rats and mice are important food items, showing the predatory potential of the species (Brothers & Skira, 1984).

Some species of burrow-nesting petrels may have been preyed on heavily by wekas. Blue petrels, common diving petrels and Antarctic prions were once abundant amongst tussocks on lower coastal slopes. The former two species are now almost confined to offshore rock stacks, and Antarctic prions now breed only in plateau herbfield.

Distribution and abundance of wekas is in part linked to that of rabbits. Destruction of suitable tall tussock grassland habitat by rabbits reduces suitable breeding habitat for wekas. Concentration of wekas in the northern half of the island was considered by Brothers & Skira (1984) to reflect the large decline of rabbit numbers in the area.

The Tasmanian Department of Parks, Wildlife and Heritage, Macquarie Island's controlling authority, has recently had a policy of eradication of wekas by shooting and trapping. Weka numbers are now very low although studies show that wekas have become more secretive in their habits and are now rarely heard calling.

Domestic animals

Domestic animals have been present on Macquarie Island from time to time as detailed in Tables 8.2 and 9.6. All were associated with sealers, oil-gatherers, the AAE or subsequent ANARE expeditions. No domestic animals have been present on the island since 1970. Mallards pose a slight problem. They are either self-introduced or deliberately introduced. We have been unable to determine the species of duck supplied to the meteorological station on the island early this century and used as food by early ANARE expeditions to the island.

Invertebrates

The only invertebrate deliberately introduced to Macquarie Island is the European rabbit flea, ecologically important as the chosen vector of myxoma virus. The potential for further accidental introductions of invertebrate species remains high.

Several invertebrate species have reached the island in recent years but there is no evidence as yet to suggest that any has become established. Two moth species not previously recorded from the island *(Agrotis ipsilon aneituma* and *Ephastia* sp.) have been caught. The collection dates of these moths strongly suggest introduction with foodstuffs during annual resupply of the ANARE base by ship and also with food supplied to the base by airdrop during the year. European wasps *(Vespula germanica)* have been found amongst bundles of building timber. There is no evidence for continued survival of any of these insects on the island. An obvious risk exists of accidental introduction of insect species to the island during normal resupply of the ANARE base.

Greenslade (1987) studied exotic invertebrates in glasshouses at Davis (Antarctica) and on Macquarie Island. She found five such species, in breeding populations, in glasshouses of the ANARE station on Macquarie Island. They had survived several months' neglect of the glasshouses but established large populations as soon as glasshouse cultivation resumed.

The risk of accidental introduction of invertebrate species to subantarctic islands, or spread of invertebrate species from one such island to another, is well illustrated in studies by Burn (1982). He described the possible introduction, associated with transfer of vascular plants, of two invertebrate species from the Falkland Islands to the South Shetland and South Orkney Islands. Invertebrate predators of Brassicaceae, such as the diamond-back moth *(Plutella xylostella)*, have recently been found on Marion Island, presumably introduced from South Africa via infected vegetables. Close monitoring of the insects' effects on the population of *Pringlea antiscorbutica* ('Kerguelen cabbage') is now under way (Watkins & Cooper, 1986).

Introduced plants

A classificatory system for alien plant species used by Walton & Smith (1973) on South Georgia and Meurk (1977) on Campbell Island includes the following categories:

(a) *transient aliens* – species surviving as one or a few individuals in a particular locality for one or two years;

(b) *persistent aliens* – species surviving as one or many individuals in a particular locality for many years;

(c) *naturalised aliens* – species with one or more populations surviving in one or more localities for many years and, spreading, successfully competing with or displacing native plants.

The flora of almost every subantarctic island has been altered by introduction of alien plant species (Table 12.2). A surprising number of

Table 12.2. *Numbers of indigenous and alien vascular plants on subantarctic islands.*

Island	Indigenous species	Naturalised alien species	Transient alien species	Reference
Macquarie	45	3	53[a]	(1), (2)
Heard	10	1	0	(3), (4)
Macdonald	5	0	0	
Kerguelen	28	3	?	(5)
Marion	24	10	4	(6), (7)
Prince Edward	21	2	0	(7)
South Georgia	23	19	45	(8)
Ile de la Possession (Crozet)	19	34	11	(9), (10)

[a] Approximately 50 species of transient aliens experimentally introduced and subsequently removed.

(1) Seppelt, Copson & Brown (1984). (2) Jenkin, Johnstone & Copson (1982). (3) Hughes (1987). (4) Scott (1989). (5) Cour (1958, 1959). (6) Smith (1987a). (7) Gremmen (1981). (8) Headland (1984). (9) Davies & Greene (1976). (10) Bell (1982).

plant species has been intentionally introduced for experimental purposes, or with the idea of 'improving' or 'adding to' local floras. Most such purposely introduced plants have proved to be transient aliens. Most naturalised aliens appear to have been accidental introductions.

Propagules of many woody plant species have been taken to Macquarie Island either for attempted deliberate introduction to the island or for experimental purposes. Such experimental introductions began in the late nineteenth century (Chapter 2). So far propagules of 39 such species have been taken to the island. No deliberate attempt at introduction of a plant species has succeeded (Jenkin, Johnstone & Copson, 1982). Plants taken to the island in the course of experiments have all been removed, most fitting into the category of transient aliens. Plants of *Calluna vulgaris* (heather), *Cotoneaster simonsii* and *Escallonia macrantha* survived in a sheltered locality for seven years as persistent aliens. All deliberately introduced plants have been removed.

A number of agricultural plant species has been introduced, some to be cultivated as vegetables in glasshouses at the ANARE station, some as stockfeed when domestic animals were present or as rabbit feed (for baiting traps) and some as experimental plants. None has become established outside glasshouses although oats, used as bait in rabbit traps (and at one stage for feeding horses (Table 9.6) grew and flowered near the ANARE base but did not set seed.

Three vascular plant species are generally considered as naturalised aliens on Macquarie Island: *Poa annua, Cerastium fontanum* and *Stellaria media.* They have been present on the island since at least the late nineteenth century and are presumed to have been introduced by sealers (Jenkin, Johnstone & Copson, 1982). All three are well established. *Poa annua* colonises disturbed ground and is ecologically important in such habitats. Moisture availability, nutrient levels and shading by overtopping species (e.g. *Poa foliosa*) may be limiting factors in its distribution. *Cerastium fontanum* is widespread but, like the more restricted *Stellaria media,* does not form dense stands.

Three other vascular species (*Corybas macranthus, Rumex crispus* and *Anthoxanthum odoratum*) have at various times been assumed to be recently introduced (Jenkin, Johnstone & Copson, 1982). *Corybas macranthus,* the only orchid yet found in the subantarctic, is an inconspicuous plant now known from many localities on Macquarie Island and is clearly not an introduced species. *Rumex crispus* is known from one locality only, north of Bauer Bay. Two clumps of the species are present. There is no justification for claims that *Rumex crispus* was introduced as a result of human activities. It is more likely to have reached the island naturally on the feet or body of a bird.

Anthoxanthum odoratum was found at the edge of a walking track near Sandy Bay. The association between the first record of the species from the island and the fact that it grew beside a well-used foot track was used to decide that the species had been introduced by human agency. All known plants of the species have been removed, making it, perforce, a transient alien. It may be, however, that other plants of *Anthoxanthum* occur elsewhere on the island. Since 1981, one plant of *Sonchus oleraceus* (milk thistle) has been found at the ANARE station and several apple seedlings close to a field hut. All these were removed, making them transient aliens.

The moss *Funaria hygrometrica* was probably introduced to Macquarie Island by humans. Plants were found beside a powerhouse and the species has been found in only that one locality. The moss plants bore capsules when discovered but are now largely overgrown by another moss (*Pottia heimii*) and vascular plants (*Cotula plumosa* and *Callitriche antarctica*). Continued survival of *Funaria hygrometrica* on Macquarie is doubtful and the species is perhaps a typical transient alien.

Detailed botanical exploration of Macquarie Island has really only just begun. For instance, a species of *Lycopodium* occurs on the island and was once listed as rare and endangered (Copson, 1984). Plants of the species have now been found at many localities on the island. Small plants such as *Lycopodium australianum, Galium antarcticum* (another recent find) and *Corybas macranthus* tend to be overlooked. There may be more to the island's flora than immediately meets the eye. As more botanical exploration is done it is likely that further additions of the island's flora will be made. In some cases it will be necessary to try to determine whether a species has been introduced by human agency or has spread to the island by natural means. Criteria for deciding such questions need to be developed. The assumption that any species found on the island for the first time is 'introduced' needs to be questioned. The subantarctic islands may have been geographically isolated for millennia but geographical isolation does not imply total biological isolation. The floras and faunas of the southern oceanic islands are too similar to have developed independently from each other. Plant propagules must have reached each island from distant sources over the centuries and many southern seabirds and seals recognise no geographical boundaries.

It is clear that Macquarie Island is a very young island, having emerged above sea-level relatively recently. It serves as a natural laboratory for the study of long-distance transport of plant and animal species.

Conservation and management strategies

Walton (1986), reporting on a joint SCAR/IUCN sponsored workshop at Paimpont, France, on the biological basis for conservation of subantarctic islands, stated:

> The principal objectives of conservation for these islands must be to maintain and protect their indigenous flora and fauna in natural associations, both by active management and by all necessary legal instruments. Policies, legislation and operational practices should be developed to constrain modification of the ecosystems, and encourage and promote a greater scientific understanding and public awareness of the importance of these islands.

Macquarie Island is, in fact, well protected by Tasmanian State legislation. A management plan is required to be prepared under this legislation and is being prepared by the Tasmanian Department of Parks, Wildlife and Heritage.

Management options for Macquarie Island can be discussed in the light of a number of recommendations formulated at the Paimpont Workshop in September 1986 (Walton, 1986).

Recommendation 1

The severity of the impact of introduced plants and animals should be assessed, appropriate control measures instituted to minimize change, and monitoring to assess ecosystem recovery.

Elsewhere in this volume the impact of alien species on the island's ecosystem has been discussed. While the effects of cats and rabbits are well documented, little is known of the interrelationships between introduced predators (e.g. cats) and native scavengers (e.g. skuas) or the full implications of removing all cats but not rabbits, or all rabbits but not cats.

Management policies of the Tasmanian Department of Parks, Wildlife and Heritage are directed firmly towards eradication of feral cats, wekas and minimising the rabbit population. Monitoring the effects of reducing numbers of these animals, as distinct from monitoring the numbers of animals themselves, is not being carried out although there have been some attempts to assess the effect of reduction in rabbit numbers on vegetation recovery.

Recommendation 2

All islands should be protected from any new accidental introductions . . . and appropriate quarantine measures implemented.

Compared to some subantarctic islands, there have been few introductions of alien species to Macquarie Island (Tables 12.1, 12.2). Quarantine restrictions have generally been implemented between Macquarie Island and Australia but not the reverse. Foreign vessels, particularly those carrying tourists, are required to obtain quarantine clearance on behalf of the Federal Department of Health prior to visiting Macquarie Island. However, the potential for accidental or deliberate introduction of alien species to the island remains high, as the recent introduction of two moth species and a wasp (see above) indicate. There have been illicit attempts to introduce peat for cultivation of vegetables in the station glasshouses. Mushroom compost has also been taken to the island. Each year there are several airdrops of mail and supplies to the island, supplementary to the normal resupply visits by ships. This practice poses even greater threats to the safe quarantine of all goods being sent to the island. Physical or biological control of an alien organism accidentally or deliberately introduced may be at best difficult, at worst impossible.

Recommendation 3

Maritime buffer zones should be instituted (where practicable under extant legislation) to provide some limited protection to food resources of the marine fauna (birds, seals).

The Macquarie Island Nature Reserve comprises those areas of Crown Land being known as Macquarie Island, Bishop and Clerk Islands, and Judge and Clerk Islands, and also includes adjacent offshore islets, rocks, and reefs and extends in all cases to low water mark. No maritime buffer zone is included in the Reserve.

Inshore pollution or commercial harvesting of fish stocks would have a significant impact on breeding seabirds and seals. Under existing Tasmanian legislation a Three Mile Economic Exclusion Zone could be proclaimed but so far has not.

Recommendation 4

Conservation policies and plans should be formulated and implemented, taking full cognisance of the impact of Man on the natural ecosystem. Scientific research and monitoring programs necessary for conservation management should be implemented.

Potentially endangered species may be protected by special regulations. The low numbers of breeding wandering albatross have been given some measure of protection by prohibition of general entry into their major breeding area near the south-west corner of the island. This prohibition extends from 7 December to 31 March each summer and runs for a trial period of five years from 1985.

It has been proposed that the fur seal breeding territories at the north end of the island should be similarly restricted during the pupping season from mid-November to about mid-January each year.

The Bishop and Clerk Islands and Judge and Clerk Islands are both subject to restricted entry to prevent transfer to them of organisms, indigenous or alien, and disturbance of any breeding populations of animals.

Several plant species, such as *Lycopodium australianum* and *Galium antarcticum,* are at present protected from collection. When first discovered the orchid, *Corybas macranthus,* was similarly protected. There is provision within the Tasmanian National Parks and Wildlife Act to add to or delete from a list of protected plant and animal species.

Recommendation 5

Island ecosystems as well as sites of specific interest should be accorded special legal protection.

In 1933, Macquarie Island was declared a Nature Sanctuary under the Tasmanian Animals and Birds Protection Act of 1928. In 1970, it was designated a State Reserve under the Tasmanian Parks and Wildlife Act and gazetted as such in 1972. Under the same Act it was declared a Nature Reserve in 1978, having been given Biosphere Reserve status in 1977 by the International Union for Conservation of Nature and Natural Resources (IUCN). IUCN has given to the island both Biosphere and Scientific or Strict Nature Reserve Status.

As a result of the protection given by Tasmanian legislation and IUCN classifications, all wildlife and natural features are protected. This includes the entire 12 785 hectares of the island and attendant offshore islets but extends only to low-water mark.

Access to the island is by permit only, issued by the Director, Tasmanian Department of Parks, Wildlife and Heritage.

Recommendation 6

Control of stations, logistics and scientific programs should ensure minimal impact on the natural ecosystems.

The ANARE station is situated on a narrow low isthmus near the northern end of the island. Six field refuges are located around the coastline. Regulations currently in force, or proposed under the new plan of management for the island, exert control over the station operation, rebuilding activities, field hut use, disposal of, or removal to Australia of, wastes arising from station occupancy or field hut use, vehicular operations (including helicopter or boat activities), and field scientific or recreational activities.

Recommendation 7

Historical sites and artefacts of shipwreck, sealing, whaling or other human activities should be mapped, documented and conserved as far as possible.

There are many recorded shipwrecks around the island (Cumpston, 1968). Many of the ships' timbers were used as a source of firewood or to build refuges by the early sealers. Relics of the sealing era have all but disappeared, the most obvious remaining relics being digesters, boilers, trypots, and a few timbers and barrel staves. An effort has been made to document as far as possible all remaining pieces of shipwreck and evidence of sealing activities in order that adequate protection may be afforded.

Recommendation 8
Education in conservation objectives for the island should be provided.

While adequate education is now provided for official and approved visits to the island, the importance of strict quarantine procedures cannot be emphasised strongly enough.

Concluding remarks
Holdgate (1970), reviewing the conservation requirements of Antarctica and the Southern Ocean, outlined three major objectives:
1. the management of natural resources for the benefit of mankind (resource conservation);
2. the protection of animal and plant species and of samples of the natural ecological systems they compose (wildlife conservation);
3. the protection of visually important features of the landscape for aesthetic reasons (conservation of amenity).

The management plan for Macquarie Island must aim at protection of the terrestrial ecosystem and near-shore waters. Declaration of a Marine Park is faced with considerable difficulty but would provide some measure of protection for the feeding grounds of much of the island's wildlife.

Whether conservation measures currently in force or proposed will be successful in preserving the ecosystem of this beautiful and fascinating island only time will tell.

References

Adamson, D. A., Selkirk, P. M. & Colhoun, E. A. (1988). Landforms of aeolian, tectonic and marine origin in the Bauer Bay – Sandy Bay region of subantarctic Macquarie Island. *Papers and Proceedings of the Royal Society of Tasmania*, **122**(1), 65–82

Adamson, D. A., Whetton, P. & Selkirk, P. M. (1988). An analysis of air temperature records for Macquarie Island: decadal warming, ENSO cooling and southern hemisphere circulation patterns. *Papers and Proceedings of the Royal Society of Tasmania*, **122**(1), 107–12

Ainsworth, G. F., Power, H. & Tulloch, A. C. (1929). Tabulated and reduced records of the Macquarie Island Station. *Australasian Antarctic Expedition 1911–1914, Scientific Reports, Series B, Volume III, Meteorology*. 473 pp.

Allison, I. F. & Keage, P. L. (1986). Recent changes in the glaciers of Heard Island. *Polar Record*, **23**(144), 255–71

Anderson, V. G. (1941). The origin of the dissolved inorganic solids in natural waters with special reference to the O'Shannassy River Catchment, Victoria. *Journal and Proceedings of the Australian Chemical Institute*, **8**, 130–50

Anon. (1987). *Macquarie Island Nature Reserve Visitor's Handbook*. Sandy Bay, Tasmania: Tasmanian National Parks and Wildlife Service. 48 pp.

Ashton, D. H. (1965). Regeneration pattern of *Poa foliosa* Hook.f. on Macquarie Island. *Proceedings of the Royal Society of Victoria*, **79**, 215–33

Ashton, D. H. & Gill, A. M. (1965). Pattern and process in a Macquarie Island feldmark. *Proceedings of the Royal Society of Victoria*, **79**, 235–45

Bailey, A. M. & Sorensen, J. H. (1962). Sub-Antarctic Campbell Island. *Denver Museum of Natural History, Proceedings*, **10**, 305 pp

Banghar, A. R. & Sykes, L. R. (1969). Focal mechanisms of earthquakes in the Indian Ocean and adjacent regions. *Journal of Geophysical Research*, **74**(2), 632–49

Barrat, A., Barre, H. & Mougin, J. L. (1976). Données écologiques sur les grands albatros *Diomedea exulans* a l'île de la Possession (Archipel Crozet). *L'oiseau et la révue française d'ornithologie*, **46**(2), 143–55

Barrow, C. J. (1978). Post-glacial pollen diagrams from South Georgia (sub-antarctic) and West Falkland Island (South Atlantic). *Journal of Biogeography*, **5**, 251–74

Bartlett, J. K. & Vitt, D. H. (1986). A survey of species of the genus *Blindia* (Bryopsida, Seligeriaceae). *New Zealand Journal of Botany*, **24**, 203–46

Beadle, N. C. W. & Costin, A. B. (1952). Ecological classification and nomenclature. *Proceedings of the Linnean Society of New South Wales,* **77**, 61–82

Bell, B. G. (1982). Notes on the alien vascular flora of l'île de la Possession, Iles Crozet. *Comité National Français des Recherches Antarctiques,* **51**, 325–31

Bellair, N. & Delibrias, G. (1967). Variations climatiques durant le dernier millénaire aux îles Kerguelen. *Comptes Rendus des Séances de l'Académie des Sciences, Paris,* **264D**, 2085–8

Bennett, I. (1971). *Shores of Macquarie Island.* Adelaide: Rigby. 69 pp

Bergstrom, D. M. (1985). *The Holocene vegetation history of Green Gorge, Macquarie Island.* M.Sc. thesis, Macquarie University, Sydney. 119 pp

Bergstrom, D. M. (1986). An atlas of seeds and fruits from Macquarie Island. *Proceedings of the Linnean Society of New South Wales,* **109**(2), 69–90

Bergstrom, D. M. & Selkirk, P. M. (1987). Reproduction and dispersal of mosses on Macquarie Island. *Symposia Biologica Hungarica,* **35**, 247–57

Berkery, B. M. & Prichard, A. (1987). Survey control for 1 : 25,000 mapping. *1986–7 Australian Antarctic Research Program. Initial Field Reports,* 151–4

Berry, R. J. & Jakobson, M. E. (1975). Adaptation and adaptability in wild living house mice *Mus musculus. Journal of Zoology (London),* **176**(3), 391–402

Berry, R. J. & Peters, J. (1975). Macquarie Island house mice: a genetical isolate on a subantarctic island. *Journal of Zoology (London),* **176**(3), 375–89

Bliss, L. C. (1979). Vascular plant vegetation of the southern circumpolar region in relation to antarctic, alpine and arctic vegetation. *Canadian Journal of Botany,* **57**(20), 2167–78

Bonner, W. N. (1964). Population increase in the fur seal, *Arctocephalus tropicalis gazella,* at South Georgia. In *Antarctic Biology,* ed. R. Carrick, M. W. Holdgate & J. Prévost, pp. 433–43. Paris: Hermann

Bonner, W. N. (1968). The fur seal of South Georgia. *British Antarctic Survey, Scientific Report,* **56**, 1–81

Bonner, W. N. (1976). The status of the Antarctic fur seal *Arctocephalus gazella. FAO Advisory Committee on Marine Resources Research, Scientific Consultation on Marine Mammals.* Bergen: Norway. ACAMRR/MM/SC/50

Brothers, N. P. (1984). Breeding, distribution and status of burrow-nesting petrels at Macquarie Island. *Australian Wildlife Research,* **11**, 113–31

Brothers, N. P. (1985). Breeding, biology, diet and morphometrics of the King Shag, *Phalacrocorax albiventer purpurascens* at Macquarie Island. *Australian Wildlife Research,* **12**, 81–94

Brothers, N. P., Eberhard, I. E., Copson, G. R. & Skira, I. J. (1982). Control of rabbits *Oryctolagus cuniculus* on Macquarie Island Australia by myxomatosis. *Australian Wildlife Research,* **9**(3), 477–85

Brothers, N. P. & Skira, I. J. (1984). The weka on Macquarie Island. *Notornis,* **31**, 145–54

Brothers, N. P., Skira, I. J. & Copson, G. R. (1985). Biology of the feral cat, *Felis catus* (L.), on Macquarie Island. *Australian Wildlife Research,* **12**, 425–36

Brown, M. J., Jenkin, J. F., Brothers, N. P. & Copson, G. R. (1978). *Corybas macranthus* (Hook.f.) Reichb.f. (Orchidaceae), a new record for Macquarie Island. *New Zealand Journal of Botany,* **16**(3), 405–7

Bryden, M. M. (1964). Insulating capacity of the subcutaneous fat of the southern elephant seal. *Nature*, **203**, 1299–1300

Bryden, M. M. (1968a). Lactation and suckling in relation to early growth of the southern elephant seal, *Mirounga leonina* (L.). *Australian Journal of Zoology*, **16**, 739–47

Bryden, M. M. (1968b). Control of growth in two populations of elephant seals. *Nature*, **217**, 1106–8

Bryden, M. M. (1969a). Relative growth of the major body components of the southern elephant seal, *Mirounga leonina* (L.). *Australian Journal of Zoology*, **17**, 153–77

Bryden, M. M. (1969b). Growth of the southern elephant seal, *Mirounga leonina* (Linn.). *Growth*, **33**, 69–82

Bryden, M. M. (1972). Body size and composition of elephant seals (*Mirounga leonina*): absolute measurements and estimates from bone dimensions. *Journal of Zoology (London)*, **167**, 265–76

Bryden, M. M. (1983).In *The Australian Museum Complete Book of Australian Mammals*, ed. R. Strahan, pp.467–9. Sydney: Angus & Robertson

Bryden, M. M. (1988). Southern elephant seals as subjects for physiological research. *Papers and Proceedings of the Royal Society of Tasmania*, **122**(1), 153–7

Bryden, M. M., Griffiths, D. J., Kennaway, D. J. & Ledingham, J. (1986). The pineal gland is very large and active in newborn Antarctic seals. *Experientia*, **42**, 564–6

Buckney, R. T. & Tyler, P. A. (1974). Reconnaissance limnology of sub-antarctic islands. II. Additional features of the chemistry of Macquarie Island lakes and tarns. *Australian Journal of Marine and Freshwater Research*, **25**(1), 89–95

Budd, G. M. (1970). Rapid population increase in the Kerguelen fur seal, *Arctocephalus tropicalis gazella*, at Heard Island. *Mammalia*, **34**, 410–14

Budd, G. M. & Downes, M. C. (1969). Population increase and breeding in the Kerguelen fur seal, *Arctocephalus tropicalis gazella*, at Heard Island. *Mammalia*, **33**, 58–67

Bull, P. C. (1960). Parasites of the European rabbit, *Oryctolagus cuniculus* (L.), on some subantarctic islands. *New Zealand Journal of Science*, **3**, 258–73

Bunt, J. (1956). Living and fossil pollen from Macquarie Island. *Nature*, **177**, 339

Bunt, J. S. (1954a). The soil-inhabiting nematodes of Macquarie Island. *Australian Journal of Zoology*, **2**, 264–74

Bunt, J. S. (1954b). A comparative account of the terrestrial diatoms of Macquarie Island. *Proceedings of the Linnean Society of New South Wales*, **79**(1–2), 34–57

Bunt, J. S. (1954c). Notes on the bacteria belonging to the Rhodobacteriineae Breed, Murray and Hitchens, and the Chlamydobacteriales Buchanan occurring at Macquarie Island. *Proceedings of the Linnean Society of New South Wales*, **79**(3–4), 63–4

Bunt, J. S. (1955). A note on the faecal flora of some Antarctic birds and mammals at Macquarie Island. *Proceedings of the Linnean Society of New South Wales*, **80**(1), 44–6

Bunt, J. S. (1965). Observations on the fungi of Macquarie Island. *ANARE Scientific Reports, Series B (II) Botany*. 22 pp

Burling, R. W. (1960). Currents in the Southern New Zealand region.
   1 : 3,394,000. *New Zealand Oceanographic Institute Chart, Miscellaneous
   Series 1* (to accompany New Zealand Oceanographic Institute Memoir 10,
   Chart 1). Wellington: Department of Scientific and Industrial Research

Burn, A. J. (1982). A cautionary tale – two recent introductions to the
   maritime Antarctic. *Colloque sur les écosystèmes subantarctique, 1981,
   Paimpont. Comité National Français des Recherches Antarctiques*, **51**, 521

Burton, H. (1986). Substantial decline in the numbers of Southern Elephant
   Seals at Heard Island. *Tasmanian Naturalist*, **86**, 4–8

Burton, R. W. (1968). Breeding biology of the Brown Skua, *Catharacta skua
   lonnbergi* (Mathews) at Signy Island, South Orkney Islands. *British
   Antarctic Survey, Bulletin*, **15**, 9–28

Butler, R. F., Banerjee, S. K. & Stout, J. H. (1975). Magnetic properties of
   oceanic pillow basalts from Macquarie Island. *Nature*, **257**, 302–3

Buynitskiy, V. K. H. (1974). Nature of the Antarctic Convergence. *Soviet
   Antarctic Expedition Information Bulletin*, **89**, 74–7 (in Russian). (Published
   in English as *Soviet Antarctic Expedition Information Bulletin, 1972–6*, **8**,
   622–4)

Calder, D. M. (1973). The effect of temperature on growth and dry weight
   distribution of populations of *Poa annua* L. In *UNESCO. Plant response to
   climatic factors. Proceedings, Uppsala Symposium, 1970. (Ecology and
   Conservation*, 5),145–52

Campbell, S. (1949). Australian aims in the Antarctic. *Polar Record*, **5**, 317–23

Carrick, R. (1972). Population ecology of the Australian black-backed magpie,
   royal penguin and silver gull. In *Population ecology of migratory birds: a
   symposium. US Department of the Interior. Wildlife Research Report*, **2**, 41–
   99

Carrick, R., Csordas, S. E. & Ingham, S. E. (1962). Studies on the southern
   elephant seal, *Mirounga leonina* (L.). IV. Breeding and development.
   CSIRO Wildlife Research, **7**, 161–97

Carrick, R., Csordas, S. E., Ingham, S. E. & Keith, K. (1962). Studies on the
   southern elephant seal, *Mirounga leonina* (L.). III. The annual cycle in
   relation to age and sex. *CSIRO Wildlife Research*, **7**, 119–60

Carrick, R. & Ingham, S. E. (1960). Ecological studies of the southern
   elephant seal, *Mirounga leonina* (L.) at Macquarie Island and Heard Island.
   Mammalia, **24**(3), 325–42

Carrick, R. & Ingham, S. E. (1962a). Studies on the Southern Elephant Seal,
   Mirounga leonina (L.). I. Introduction to the series. *CSIRO Wildlife
   Research*, **7**, 89–101

Carrick, R. & Ingham, S. E. (1962b). Studies on the Southern Elephant Seal,
   Mirounga leonina (L.). V. Population dynamics and utilization. *CSIRO
   Wildlife Research*, **7**, 198–206

Carrick, R. & Ingham, S. E. (1967). Antarctic sea-birds as subjects for
   ecological research. *JARE Scientific Reports, Special Issue Number 1.
   Proceedings of Symposium on Pacific–Antarctic Sciences, Tokyo*, 151–84

Carrick, R. & Ingham, S. E. (1970). Ecology and population dynamics of
   Antarctic sea birds. In *Antarctic Ecology*, ed. M. W. Holdgate, pp. 505–25.
   London: Academic Press

Chappell, J. (1983). A revised sea-level curve for the last 300,000 years from
   Papua New Guinea. *Search*, **14**(3–4), 99–101

Chappell, J. & Shackleton, N. J. (1986). Oxygen isotopes and sea level. *Nature*, **324**, 137–40

Cheeseman, T. F. (1919). The vascular flora of Macquarie Island. *Australasian Antarctic Expedition 1911–1914. Scientific Reports. Series C. Zoology and Botany*. Volume VII, Part 3. 63 pp

Christodoulou, C., Griffin, B. J. & Foden, J. (1984). The geology of Macquarie Island. *ANARE Research Notes*, **21**. 15 pp

Clark, M. R. & Dingwall, P. R. (1985). *Conservation of islands in the Southern Ocean. A review of the protected areas of Insulantarctica*. Gland, Switzerland: IUCN. 188 pp

Clifford, H. T. (1953). The mosses of Macquarie Island and Heard Island. *ANARE Reports, Series B, II, Botany*. 14 pp

Cocker, J. D., Griffin, B. J. & Muehlenbachs, K. (1982). Oxygen and carbon isotope evidence for sea-water hydrothermal alteration of the Macquarie Island ophiolite. *Earth and Planetary Science Letters*, 61, 112–22

Colhoun, E. A. & Goede, A. (1973). Fossil penguin bones, 14C dates and the raised marine terrace of Macquarie Island: some comments. *Search*, **4**(11–12), 499–501

Colhoun, E. A. & Goede, A. (1974). A reconnaissance survey of the glaciation of Macquarie Island. *Papers and Proceedings of the Royal Society of Tasmania*, **108**, 1–19

Colhoun, E. A. & Peterson, J. A. (1986). Quaternary landscape evolution and the cryosphere: research progress from Sahul to Australian Antarctica. *Australian Geographical Studies*, **24**, 145–67

Condy, P. R. (1978). The distribution and abundance of southern elephant seals (*Mirounga leonina* (Linn.)) on the Prince Edward Islands. *South African Journal of Antarctic Research*, **8**, 42–8

Condy, P. R. (1979). Elephant seals on Marion Island. *African Wildlife*, **33**(1), 36–7

Conroy, J. W. H. (1972). Ecological aspects of the biology of the giant petrel. *Macronectes giganteus* Gmelin in the maritime Antarctic. *British Antarctic Survey, Scientific Reports*, **75**, 1–74

Conroy, J. W. H. & Twelves, E. L. (1972). Diving depths of the gentoo penguin (*Pygoscelis papua*) and blue-eyed shag (*Phalacrocorax atriceps*) from the South Orkney Islands. *British Antarctic Survey, Bulletin*, **30**, 106–8

Copson, G. R. (1984). An annotated atlas of the vascular flora of Macquarie Island. *ANARE Research Notes*, **18**. 70 pp

Copson, G. R. (1986). The diets of the introduced rodents *Mus musculus* L. and *Rattus rattus* L. on subantarctic Macquarie Island. *Australian Wildlife Research*, **13**, 441–5

Copson, G. R. (1988). The status of the Black-browed and Grey-headed Albatrosses on Macquarie Island. *Papers and Proceedings of the Royal Society of Tasmania*, **122**(1),137–41

Copson, G. R., Brothers, N. P. & Skira, I. J. (1981). Distribution and abundance of the rabbit, *Oryctolagus cuniculus* (L.), at subantarctic Macquarie Island. *Australian Wildlife Research*, **8**, 597–611

Copson, G. R. & Leaman, E. G. (1981). *Rumex crispus* L. (Polygonaceae) – a new record for Macquarie Island. *New Zealand Journal of Botany*, **19**, 401–4

Copson, G. R. & Rounsevell, D. E. (1987). The abundance of Royal penguins

(*Eudyptes schlegeli*, Finsch) breeding at Macquarie Island. *ANARE Research Notes*,**41**. 11 pp

Costin, A. B. & Moore, D. M. (1960). The effects of rabbit grazing on the grasslands of Macquarie Island. *Journal of Ecology*, **48**, 729–32

Cour, P. (1958).À propos de la flore de l'Archipel de Kerguelen. *Terres Australes et Antarctiques françaises*, nos. 4 and 5, 10–32

Cour, P. (1959). Flore et végétation de l'Archipel de Kerguelen. *Terres Australes et Antarctiques françaises*, nos. 8 and 9, 3–40

Crohn, P. W. (1986). The geology and geomorphology of Macquarie Island with special emphasis on heavy metal trace element distribution. *ANARE Research Notes*, **39**. 28 pp

Croome, R. L. (1972). Preliminary investigations on the nitrogen status of the Marion Island ecosystem. Marion and Prince Edward Island. Biological and Geological Research Report of the 1971–72 Expedition. Institute for Environmental Sciences, University of the Orange Free State, Bloemfontein, South Africa. 81 pp (unpublished)

Croome, R. (1984). Limnological studies on Macquarie Island. *Tasmanian Naturalist*, **78**, 26–7

Croxall, J. P. (1979). Distribution and population changes in the Wandering Albatross *Diomedea exulans* at South Georgia. *Ardea*, **67**, 15–21

Croxall, J. P. (1982). Aspects of the population demography of Antarctic and subantarctic sea birds. *Colloque sur les écosystèmes subantarctiques, 1981, Paimpont. Comité National Français des Recherches Antarctiques*, **51**, 479–88

Croxall, J. P. & Prince, P. A. (1979). Antarctic seabird and seal monitoring studies. *Polar Record*, **19**, 573–95

Croxall, J. P. & Prince, P. A. (1980). The food of Gentoo Penguins *Pygoscelis papua* and Macaroni Penguins *Eudyptes chrysolophus* at South Georgia. *Ibis*, **122**, 245–53

Csordas, S. E. (1958). Breeding of the fur seal, *Arctocephalus forsteri* Lesson, at Macquarie Island. *Australian Journal of Science*, **21**, 87–8

Csordas, S. (1985). Domestic animals on Macquarie Island. *Aurora*, **18**, 23–6

Csordas, S. E. & Ingham, S. E. (1965). The New Zealand Fur seal, *Arctocephalus forsteri* (Lesson), at Macquarie Island, 1949–1964. *CSIRO Wildlife Research*, **10**, 83–99

Cumpston, J. S. (1968). *Macquarie Island. ANARE Scientific Reports, Series A (1), Narrative.* Melbourne: Antarctic Division, Department of External Affairs. 380 pp

Davies, L. & Greene, S. W. (1976). Notes sur la végétation de l'île de la Possession (Archipel Crozet). *Comité National Français des Recherches Antarctiques*, **41**, 1–20

Dawson, E. W. (1988). The offshore fauna of Macquarie Island: history and biogeography – results from New Zealand and United States research cruises. *Papers and Proceedings of the Royal Society of Tasmania*, **122**(1), 219–32

Deacon, G. E. R. (1960). The southern cold temperate zone. In *A discussion of the biology of the southern cool temperate zone*, (leader) C. F. A. Pantin, pp. 441–7. *Proceedings of the Royal Society, London, B*, **152**

Deacon, G. E. R. (1966). Convergence, Antarctic. In *Encyclopedia of Oceanography*, ed. R. W. Fairbridge, pp. 215–19. New York: Reinhold

Dell, R. K. (1964). Marine Mollusca from Macquarie and Heard Islands. *Records of the Dominion Museum*, **4**(20), 267–301

Division of National Mapping, Australia (1971). *Macquarie Island, Tasmania.* 1 : 50 000. Canberra, Australia

Division of National Mapping, Australia (1978). *Antarctica and Adjacent Continents,* scale at 70°S, 1 : 40 000. Canberra, Australia

Division of National Mapping, Australia (1982). *Manuscript Bathymetric Map, Macquarie Island.* 1 : 50 000, Sheets 1 & 2. Canberra, Australia

Dodge, C. W. (1948). Lichens and lichen parasites. *British, Australian and New Zealand Antarctic Research Expedition 1929–1931. Reports. Series B. Zoology and Botany*, **7**, 276 pp

Dodge, C. W. (1968). Lichenological notes on the flora of the Antarctic continent and the subantarctic islands. VII, VIII. *Nova Hedwigia*, **15**, 285–332

Dodge, C. W. (1970). Lichenological notes on the flora of the Antarctic continent and subantarctic islands. IX–XI. *Nova Hedwigia* **19**, 439–502

Dodge, C. W. & Rudolph, E. D. (1955). Lichenological notes on the flora of the Antarctic continent and the subantarctic islands. II. Additions to the lichen flora of Macquarie Island. *Annals of the Missouri Botanical Garden*, **42**, 137–43

Doherty, R. L., Carley, J. G., Murray, M. D., Main, A. J., Kay, B. H. & Domrow, R. (1975). Isolation of arboviruses (Kemerovo group Sakhalin group) from *Ixodes uriae* collected at Macquarie Island, Southern Pacific Ocean. *American Journal of Tropical Medicine and Hygiene*, **24**(3), 521–6

Dorst, J. & Millon, Ph. (1964). Acclimation et conservation de la nature dans îles subantarctiques françaises. In *Antarctic Biology*, ed. R. Carrick, M. Holdgate & J. Prévost, pp. 579–88. Paris: Hermann

Downes, M. C., Ealey, E. H. M., Gwynn, A. M. & Young, P. S. (1959). The birds of Heard Island. *ANARE Scientific Reports, Series B*, 1, *Zoology*, 135 pp

Duncan, R. A. and Varne, R. (1988). The age and distribution of the igneous rocks of Macquarie Island. *Papers and Proceedings of the Royal Society of Tasmania*, **122**(1), 45–50

Edgar, E. (1966). *Luzula* in New Zealand. *New Zealand Journal of Botany*, **4**, 159–84

Edgar, E.(1986). *Poa* L. in New Zealand. *New Zealand Journal of Botany*, **24**, 425–503

Ellis, W. M., Lee, B. T. O. & Calder, D. M. (1971). A biometric analysis of populations of *Poa annua* L. *Evolution*, **25**, 29–37

Embleton, C. & King, C. A. M. (1975). *Periglacial Geomorphology.* New York: John Wiley & Sons. 203 pp

Evans, A. J. (1970). Some aspects of the ecology of a calanoid copepod, *Pseudoboeckella brevicaudata* Brady 1875, on a subantarctic island. *ANARE Scientific Reports, Series B, 1, Zoology*, 100 pp

Falla, R. A. (1937). Birds. *British, Australian and New Zealand Antarctic Research Expedition 1929–1931, Reports, Series B, Zoology and Botany*, **2**. 304 pp

Falla, R. A. (1962). Exploitation of seals, whales and penguins in New Zealand. *Proceedings of the New Zealand Ecological Society*, **9**, 34–8

Filson, R. B. (1981). Studies on Macquarie Island lichens 2: the genera

Hypogymnia, Menegazzia, Parmelia and *Psedocyphellaria. Muelleria,* **4**(4), 317–31

Filson, R. B. (1986). Studies in Macquarie Island lichens 3: the genus *Sphaerophorus. Muelleria,* **6**(3), 169–72

Filson, R. B. & Archer, A. W. (1986). Studies in Macquarie Island lichens 4: the genera *Cladia* and *Cladonia. Muelleria,*6(3), 217–35

Fischer, W. & Hureau, J. C. (eds.) (1985). FAO species identification sheets for fishery purposes. Southern Ocean (Fishing areas 48, 58 and 88) (CCAMLR Conservation Area). Prepared and published with the support of the Commission for the Conservation of Antarctic Marine Living Resources. Rome: FAO, **1**, 1–232

Fletcher, L. & Shaughnessy, P. (1984). The current status of seal populations at Macquarie Island. *Tasmanian Naturalist,* **79**, 10–13

Folland, C. K., Parker, D. E. & Kates, F. E. (1984). Worldwide marine temperature fluctuations 1856–1981. *Nature,* **310**, 670–3

Foster, T. D. (1984). The marine environment. In *Antarctic Ecology,* ed. R. M. Laws, pp. 345–72. London: Academic Press

Fressanges du Bost, D. & Segonzac, M. (1976). Note complémentaire sur le cycle reproducteur du Grand Albatros (*Diomedea exulans*) à l'île de la Possession, archipel Crozet. *Comité National Français des Recherches Antarctiques,* **40**, 53–60

Frith, W. J. (1982). *Waterfowl in Australia,* revised edition. Sydney: Angus & Robertson

Gales, N. J. & Burton, H. R. (1987). Ultrasonic measurement of blubber thickness of the Southern Elephant seal, *Mirounga leonina* (Linn.). *Australian Journal of Zoology,* **35**, 207–17

Gardner, Z. N. C. (1977). Two floating islands on a subantarctic lake. *Antarctic Division Technical Memorandum.* ANARE Publication Number 63. 25 pp

Gass, I. G. (1982). Ophiolites. *Scientific American,* **247**, 122–31

Gibbney, L. F. (1957). The seasonal reproductive cycle of the female Elephant Seal – *Mirounga leonina,* Linn. – at Heard Island. *ANARE Reports, Series B, Volume 1, Zoology.* 26 pp

Gillham, M. E. (1961). Modification of sub-antarctic flora on Macquarie Island by sea birds and sea elephants.*Proceedings of the Royal Society of Victoria,* **74**(1), 1–12

Gordon, A. L. (1967). Structure of Antarctic Waters between 20°W and 170°W. In *Antarctic Map Folio Series, Folio 6,* ed. V. C. Bushnell. New York: American Geographical Society

Gordon, A. L. (1972). On the interaction of the Antarctic Circumpolar Current and the Macquarie Ridge. In *Antarctic Oceanology,* ed. D. E. Hayes, pp. 71–8. Washington DC: American Geophysical Union. American Geophysical Union Antarctic Research Series, 19

Gordon, A. L. & Goldberg, R. D. (1970). Circumpolar characteristics of Antarctic waters. In Antarctic Map Folio Series, Folio 13, ed. V. C. Bushnell. New York: American Geographical Society

Greenslade, P. (1987). Soil fauna studies with particular reference to collembolar faunistics, densities and distributions. *1986–87 Australian Antarctic Research Program. Initial Field Reports,* 80–4

Greenslade, P. & Wise, K. A. J. (1986). Collembola of Macquarie Island. *Records of the Auckland Institute and Museum,* **23**, 67–97

Gremmen, N. J. M. (1981). The vegetation of the subantarctic islands Marion and Prince Edward. *Geobotany 3.* The Hague: Dr W. Junk. 149 pp

Gressitt, J. L., Forster, R. R., Fain, A., Smithers, C. N., Stannard, L. J. (Jr)., Eastop, V. F., Alexander, C. P., Brundin, L., Colless, D. H., Quate, L. W., Kohn, M. A., Hardy, D. E., Wirth, W. W., Dunnet, G. M., Sabrosky, C. W., Yoshimoto, C. M. & Common, I. F. B. (1962). Insects of Macquarie Island. *Pacific Insects,* **4**(4), 905–78. (A collection of separate papers issued as a single Part with an introduction to the series by J. L. Gressitt)

Griffin, B. J. (1980). Erosion and rabbits on Macquarie Island: some comments. *Papers and Proceedings of the Royal Society of Tasmania,* **114**, 81–3

Griffin, B. J. & Varne, R. (1980). The Macquarie Island ophiolite complex: mid-Tertiary oceanic lithosphere from a major ocean basin. *Chemical Geology,* **30**, 285–308

Griffiths, D., Seamark, R. F. & Bryden, M. M. (1979). Summer and winter cycles in plasma melatonin levels in the Elephant seal (*Mirounga leonina*). *Australian Journal of Biological Science,* **32**, 581–6

Griffiths, D. J. (1984a). The annual cycle of the epididymis of the elephant seal (*Mirounga leonina*) at Macquarie Island. *Journal of Zoology, London,* **203**, 181–91

Griffiths, D. J. (1984b). The annual cycle of the testis of the elephant seal (*Mirounga leonina*) at Macquarie Island. *Journal of Zoology, London,* **203**, 193–204

Griffiths, D. J. & Bryden, M. M. (1981). The annual cycle of the pineal gland of the elephant seal (*Mirounga leonina*).In *Pineal Function,* ed. C. D. Matthews & R. F. Seamark, pp. 57–66. Amsterdam: Elsevier/North Holland

Grobbelaar, J. V. (1978). Mechanisms controlling the composition of fresh waters on the sub-antarctic island Marion. *Archiv für Hydrobiologie,* **83**(2), 145–57

Grolle, R. & Seppelt, R. D. (1986). *Seppeltia,* a new leafy genus of Metzgeriales from Macquarie Island. *Journal of the Hattori Botanical Laboratory,* **60**, 275–82

Gwynn, A. M. (1953a). The egg-laying and incubation periods of Rockhopper, Macaroni and Gentoo penguins. *ANARE Reports, Series B,* 1, *Zoology.* 27 pp

Gwynn, A. M. (1953b). Notes on the fur seals at Macquarie Island and Heard Island. *ANARE Interim Reports,* **4**. 16 pp

Hall, K. (1981). Observations on the stone-banked lobes of Marion Island. *South African Journal of Science,* **77**, 129–31.

Hamilton, A. (1895). Notes on a visit to Macquarie Island. *Transactions and Proceedings of the New Zealand Institute 1894,* **27**, 559–79.

Hamilton, H. (1926). Ecological notes and illustrations of the flora of Macquarie Island. *Australasian Antarctic Expedition 1911–1914. Scientific Reports, Series C. Zoology and Botany.* Volume VII, Part 5. 10 pp + 19 plates

Harland, W. B., Cox, A. V., Llewellyn, P. G., Pickton, C. A. G., Smith, A.

G. & Walters, R. (1982). *A geologic time scale.* Cambridge: Cambridge University Press. 131 pp

Hayes, D. E. & Talwani, M. (1972). Geophysical investigation of the Macquarie Ridge complex. In *Antarctic Oceanology II The Australian–New Zealand Sector,* ed. D. E. Hayes, pp. 211–34. Washington DC: American Geophysical Union. American Geophysical Union Antarctic Research Series, 19.

Headland, R. (1984). *The Island of South Georgia.* Cambridge: Cambridge University Press. 293 pp

Hindell, M. A. (1988). The diet of the King penguin *Aptenodytes patagonicus* at Macquarie Island. *Ibis,* **130,** 193–203

Holdgate, M. W. (1966). The influence of introduced species on ecosystems of temperate oceanic islands. In *Towards a new relationship of man and nature in temperature lands. Proceedings of the 10th Technical Meeting, IUCN, Lucerne. IUCN Publication, New Series,* **9,** 151–76

Holdgate, M. W. (1970). Conservation in the Antarctic. In *Antarctic Ecology,* vol. 2, ed. M. W. Holdgate, pp. 924–45. London: Academic Press

Holdgate, M. W. & Wace, N. M. (1961). The influence of man on the floras and faunas of southern islands. *Polar Record,* **10,** 475–93

Houtman, T. J. (1965). Temperature and salinity distribution south of New Zealand. *New Zealand Oceanographic Institute Chart, Miscellaneous Series No. 10* (to accompany New Zealand Oceanographic Institute Memoir No. 36, Chart No. 1). Wellington: Department of Scientific and Industrial Research

Houtman, T. J. (1967). Water masses and fronts in the Southern Ocean south of New Zealand. *New Zealand Department of Scientific and Industrial Research Bulletin,* **174,** 1–39

Howard, P. F. (1954). ANARE bird banding and seal marking. *Victorian Naturalist,* **71,** 73–82

Hughes, J. M. R. (1986). The relations between aquatic plant communities and lake characteristics on Macquarie Island. *New Zealand Journal of Botany,* **24,** 271–8

Hughes, J. M. R. (1987). The distribution and composition of vascular plant communities on Heard Island. *Polar Biology,* **7,** 153–62

Ingham, S. E. (1984). A history of Macquarie Island biological research up to 1971. *Tasmanian Naturalist,* **78,** 3–5

Inoue, H. & Seppelt, R. D. (1985). Notes on the genus *Plagiochila* (Dum.) Dum. from subantarctic Macquarie Island. *Bulletin of the National Science Museum, Series B (Botany),* **11**(4), 119–26

Jacka, T. H. (1983). A computer data base for Antarctic sea ice extent. *ANARE Research Notes,* **13.** 54 pp

Jacka, T. H., Christou, L. & Cook, B. J. (1984). A data bank of mean monthly and annual surface temperatures for Antarctica, the Southern Ocean and the South Pacific Ocean. *ANARE Research Notes,* **22.** 97 pp

Jamieson, B. G. M. (1968). *Macquaridrilus*: a new genus of Tubificidae (Oligochaeta) from Macquarie Island. University of Queensland Papers, Department of Zoology, **3**(5), 55–69

Jenkin, J. F. (1972). Studies on plant growth in a subantarctic environment. Ph.D. thesis, University of Melbourne. 297 pp

Jenkin, J. F. (1975). Macquarie Island, Subantarctic. In *Structure and Function of Tundra Ecosystems,* ed. T. Rosswall & O. W. Heal, pp. 375–97. Stockholm: Ecological Bulletins, 20

Jenkin, J. F. & Ashton, D. H. (1970). Productivity studies on Macquarie Island vegetation. In *Antarctic Ecology,* vol. 2, ed. M. W. Holdgate, pp. 851–63. New York: Academic Press

Jenkin, J. F. & Ashton, D. H. (1979). Pattern in *Pleurophyllum* herbfields on Macquarie Island (sub-antarctic). *Australian Journal of Ecology,* **4**, 47–66

Jenkin, J. F., Johnstone, G. W. & Copson, G. R. (1982). Introduced animal and plant species on Macquarie Island. *Colloque sur les écosystèmes subantarctiques. 1981, Paimpont. Comité National Français des Recherches Antarctiques,* **51**, 301–13

Johnson, T. & Molnar, P. (1972). Focal mechanisms and plate tectonics of the Southwest Pacific. *Journal of Geophysical Research,* **77**(26), 5000–32

Johnston, G. C. (1973). Predation by southern skua on rabbits on Macquarie Island. *The Emu,* **73**(1), 25–26

Johnstone, G. W. (1977). Comparative feeding ecology of the giant petrels *Macronectes giganteus* (Gmelin) and *M. halli* (Mathews). In *Adaptations within Antarctic Ecosystems. Proceedings of the 3rd SCAR Symposium on Antarctic Biology,* ed. G. A. Llano, pp. 647–68. Washington DC: Smithsonian Institute

Johnstone, G. W. (1978). Interbreeding by *Macronectes halli* and *M. giganteus* at Macquarie Island. *The Emu,* **78**(4), 235

Johnstone, G. W. (1980). Australian islands in the Southern Ocean. *The Bird Observer,* October 1980, **586**, 85–7

Johnstone, G. W. (1985). Threats to birds on sub-antarctic islands. *International Council for Bird Preservation. Technical Publications,* **33**, 101–21

Jones, E. (1977). Ecology of the feral cat *Felis catus* (L.), (Carnivora: Felidae) on Macquarie Island. *Australian Wildlife Research,* **4**(4), 249–62

Jones, E. (1980). Survey of burrow-nesting petrels at Macquarie Island based upon remains left by predators. *Notornis,* **27**(1), 11–20

Jones, E. (1984). The feral cat on Macquarie Island. *Tasmanian Naturalist,* **79**, 16–17

Jones, E. & Skira, I. J. (1979). Breeding distribution of the Great Skua at Macquarie Island in relation to the numbers of rabbits. *The Emu,* **79**(1), 19–23

Jones, P. D., Wigley, T. M. L. & Wright, P. B. (1986). Global temperature variations between 1861 and 1984. *Nature,* **332**, 330–4

Jones, R. (1987). Participation of a prehistory scholar in ANARE activities. *1986–87 Australian Antarctic Research Program. Initial Field Reports,* 215–17

Jones, T. D. & McCue, K. R. (1988). Seismicity and tectonics of the Macquarie Ridge. *Papers and Proceedings of the Royal Society of Tasmania,* **122**(1), 51–7

Kelly, P. M., Jones, P. D., Sear, C. B., Cherry, B. S. G. & Tavakol, R. K. (1982). Variations in surface air temperatures: Part 2. Arctic regions, 1881–1980.*Monthly Weather Review,* **110**, 71–83

Kenny, R. & Haysom, N. (1962). Ecology of rocky shore organisms at Macquarie Island. *Pacific Science,* **16**(3), 245–63

Kep, S. L. (1984). A climatology of cyclogenesis, cyclone tracks and cyclolysis in the Southern Hemisphere for the period 1972–1981. *Meteorology Department University of Melbourne Publication*, **25**. Parkville: University of Melbourne

Kerr, R. A. (1983). Ophiolites: windows on which ocean crust? *Science*, **219**, 1307–9

Kerry, E. (1984). The fungal flora of Macquarie Island. *Tasmanian Naturalist*, **78**, 16–21

Kerry, E. J. & Weste, G. M. (1985). Succession in the microflora of leaves and litter of three plants from subantarctic Macquarie Island. In *Antarctic Nutrient Cycles and Food Webs. Proceedings of the Fourth Symposium on Antarctic Biology. Wilderness, South Africa*, ed. R. Siegfried, P. R. Condy & R. M. Laws, pp. 597–605. Heidelberg: Springer-Verlag

Kerry, K. R. & Colback, G. C. (1972). Follow the band! Light-mantled Sooty albatrosses on Macquarie Island. *Australian Bird Bander*, **10**(3), 61–2

Kirkpatrick, J. B. (1984). Tasmanian high mountain vegetation II – Rocky Hill and Pyramid Mountain. *Papers and Proceedings of the Royal Society of Tasmania*, **118**, 5–20

Kirkwood, J. M. (1982). A guide to the Euphausiacea of the Southern Ocean. *ANARE Research Notes*, **1**. 45 pp

Klein, J., Lerman, J. C., Daman, P. G. & Ralph, E. K. (1982). Calibration of radiocarbon dates: tables based on concensus data of the workshops on calibrating the radiocarbon timescale. *Radiocarbon*, **24**, 103–50

Klemm, M. & Hallam, N. (1988). Conceptacle development, gamete maturation and embryology of *Durvillaea antarctica* from Macquarie Island. *Papers and Proceedings of the Royal Society of Tasmania*, **122**(1), 199–210

Lambeth, A. J. (1951). Heard Island, geography and glaciology. *Journal and Proceedings of the Royal Society of New South Wales*, **84**, 92–8

Law, P. G. (1961). Australian National Antarctic Research Expeditions, 1959–60. *Polar Record*, **10**(67), 397–401

Law, P. G. (1962). Australian National Antarctic Research Expeditions, 1960–61. *Polar Record*, **11**(71), 184–7

Law, P. G. & Burstall, T. (1956). Macquarie Island. *ANARE Interim Reports*, **14**. 48 pp

Laws, R. M. (1953). The Elephant Seal (*Mirounga leonina* Linn.) I. Growth and age. *Falkland Islands Dependencies Survey Scientific Reports*, **8**. 62 pp

Laws, R.M. (1977). Seals and whales of the Southern Ocean. *Philosophical Transactions of the Royal Society, London, B*, **279**, 81–96

Ledingham, R. & Peterson, J. A. (1984). Raised beach deposits and the distribution of structural lineaments on Macquarie Island. *Papers and Proceedings of the Royal Society of Tasmania*, **118**, 223–35

Lesel, R. & Derenne, P. (1975). Introducing animals to Kerguelen. *Polar Record*, **17**, 485–94

Lewis, J. R. (1961). The littoral zone on rocky shores – a biological or physical entity? *Oikos*, **12**, 280–301

Lewis, J. R. (1964). *The ecology of rocky shores*. London: English Universities Press, 323 pp

Lightowlers, P. J. (1986). Taxonomy and distribution of the subantarctic species of *Tortula*. *Journal of Bryology*, **14**(2), 281–95

Ling, J. K. (1965). Hair growth and moulting in the Southern Elephant seal, *Mirounga leonina* (Linn.). In *Biology of the Skin and Hair Growth.*

Proceedings of a symposium held at Canberra, Australia, 22–28 August, 1964, 525–44. Sydney: Angus and Robertson

Loewe, F. (1957). A note on sea water temperatures on Macquarie Island. *Australian Meteorological Magazine*, **19**, 60–1

Löffler, E. (1983). Macquarie Island – eine vom Wind geprägte Naturlandschaft in der Sub-Antarktis. *Polarforschung*, **53**(1), 59–74

Löffler, E. & Sullivan, M. E. (1980). The extent of former glaciation on Macquarie Island. *Search*, **11**(7–8), 246–7

Löffler, E., Sullivan, M. E. & Gillison, A. N. (1983). Periglacial landforms on Macquarie Island, Subantarctic. *Zeitschrift für Geomorphologie N.F.*, **27**(2), 223–36

Longton, R. E. (1977). A *Nothofagus* log stranded on Candelmas Island, South Sandwich Islands. *British Antarctic Survey, Bulletin*, **45**, 148–9

Longton, R. E. (1988). Adaptations and strategies of polar bryophytes. *Botanical Journal of the Linnean Society*, **98**, 253–63

Lowry, J. K., Horning, D. S., Poore, G. C. B. & Ricker, R. W. (1978). *The Australian Museum Macquarie Island Expedition, Summer 1977–1978*. Sydney: Australian Museum Trust. 152 pp

Lugg, D. J., Johnstone, G. W. & Griffin, B. J. (1978). The outlying islands of Macquarie Island. *The Geographical Journal*, **144**, 277–87

McEvey, A. R. & Vestjens, W. J. M. (1973). Fossil penguin bones from Macquarie Island, southern ocean. *Proceedings of the Royal Society of Victoria*, **86**(2), 151–74

MacKenzie, D. (1968). The birds and seals of the Bishop and Clerk Islets, Macquarie Island. *The Emu*, **67**(4), 241–5

McLeod, I. (1986). Macquarie Island and domestic animals. *Aurora*, **5**(4), 33

Mallis, M. (1985). A qualitative investigation into scavenging of airborne sea salt over Macquarie Island. *ANARE Research Notes*, **26**. 18 pp

Mallis, M. (1988). A quantitative analysis of aerosol (salt) scavenging on Macquarie Island. *Papers and Proceedings of the Royal Society of Tasmania*, **122**(1), 121–8

Mawson, D. (1919). Macquarie Island. A sanctuary for Australasian Sub-antarctic fauna. *Proceedings of the Royal Geographical Society of Australasia. South Australian Branch 1918–1919*, 71–85

Mawson, D. (1922). Macquarie Island and its future. *Papers and Proceedings of the Royal Society of Tasmania, 1922*, 40–54. Reprinted in *Australian Zoologist*, **3**, 92–102

Mawson, D. (1943). Macquarie Island: its geography and geology. *Australasian Antarctic Expedition Scientific Reports, Series A, Volume V*. 193 pp

Mawson, P. M. (1953). Parasitic nematoda collected by the Australian National Antarctic Research Expedition: Heard Island and Macquarie Island, 1948–51. *Parasitology*, **43**, 291–7

Merilees, W. J. (1971). Bird observations – Macquarie Island, 1967. *Notornis*, **18**, 55–6

Merilees, W. (1984a). Some notes on the foods of the Dominican Gull at Macquarie Island. *Tasmanian Naturalist*, **79**, 5–6

Merilees, W. (1984b). On a mass mortality of lantern fish at Macquarie Island. *Tasmanian Naturalist*, **78**, 32

Meurk, C. D. (1977). Alien plants in Campbell Island's changing vegetation. *Mauri Ora*, **5**, 93–118

Mills, J. A. (1976). Status, mortality and movements of grey teal (*Anas gibberifrons*) in New Zealand. *New Zealand Journal of Zoology*, **3**, 261–7

Moors, P. J. (1986). Decline in numbers of Rockhopper Penguins at Campbell Island. *Polar Record*, **23**, 69–73

Morgan, I. (1988). Avian viruses on Macquarie Island. *Papers and Proceedings of the Royal Society of Tasmania*, **122**(1), 193–8

Morgan, I. R., Caple, I. W., Westbury, H. A. & Campbell, J. (1978). Disease investigations of penguins and Elephant seals on Macquarie Island. *Research Project Series*, **47**. Melbourne: Department of Agriculture. 51 pp

Mougin, J. L. (1968). Etude écologique de quatre espèces de petrels antarctiques. *L'oiseau et la revue française d'ornithologie*, Special Number **38**, 1–52

Mougin, J. L. (1969). Notes écologiques sur le pétrel de Kerguelen *Pterodroma brevirostris* de l'Ile de la Possession (Archipel Crozet). *L'oiseau et la revue française d'ornithologie*. Special Number **39**, 58–81

Mougin, J. L. (1977). Nidification à l'île Marion (46°53'S, 37°52'E) d'un Grand Albatros (*Diomedea exulans* L.) né à l'île de Possession, archipel Crozet (46°25'S, 51°45'E). *Comptes Rendus des Séances de l'Académie des Sciences, Paris, Series D*, **284**, 2277–80

Murray, M. D. (1958). Ecology of the louse *Lepidophthirus macrorhini* on the elephant seal *Mirounga leonina* (L.). *Nature*, **182**, 404–5

Murray, M. D. (1964). Ecology of the ectoparasites of seals and penguins. In *Antarctic Biology*, ed. R. Carrick, M. W. Holdgate & J. Prévost, pp. 241–5. Paris: Hermann

Murray, M. D. & Nicholls, D. G. (1965). Studies on the ectoparasites of seals and penguins. I. The ecology of the louse *Lepidophthirus macrorhini* Enderlein on the Southern Elephant Seal, *Mirounga leonina* (L.). *Australian Journal of Zoology*, **13**, 437–54

Murray, M. D. & Vestjens, W. J. M. (1967). Studies on the ectoparasites of seals and penguins. III. The distribution of the tick *Ixodes uriae* White and the flea *Parapsyllus magellanicus heardi* de Meillon on Macquarie Island. *Australian Journal of Zoology*, **15**, 715–25

Norman, F. I. (1987). The ducks of Macquarie Island. *ANARE Research Notes*, **42**. 22 pp

Norman, F. I. & Brown, R. S. (1987). Aspects of the ecology of Macquarie Island waterfowl. *1986–87 Australian Research Programs. Initial Field Reports*, 101–4

Orchard, A. E. (1986). A revision of the *Coprosma pumila* (Rubiaceae) complex in Australia, New Zealand and the subantarctic islands. *Brunonia*, **9**, 119–38

Orchard, A. E. (1989). *Azorella* Lamarek (Hydrocotylaceae) on Heard and Macquarie Islands. *Muelleria*, **7**(1), 15–20

O'Sullivan, D. (1983). Fisheries of the Southern Ocean. *Australian Fisheries*, July, 1983, 4–11

O'Sullivan, D. B., Johnstone, G. W., Kerry, K. R. & Imber, M. J. (1983). A mass stranding of squid *Martialia hyadesi* Rochebrunne & Mabille (Teuthoidea: Ommastrephidae) at Macquarie Island. *Papers and Proceedings of the Royal Society of Tasmania*, **117**, 161–3

Payne, M. R. (1977). Growth of a fur seal population. *Philosophical Transactions of the Royal Society, London, B*, **279**, 67–79

Peterson, J. A. (1975). The morphology of Major Lake, Macquarie Island. *Australian Society for Limnology Bulletin,* **6,** 17–26

Peterson, J. A. (1987). Quaternary history of Macquarie Island. *1986–87 Australian Antarctic Research Program. Initial Field Reports,* Antarctic Division, Australia, 74–6

Peterson, J. A. & Scott, J. J. (1988). Interrelationships between wind exposure, vegetation, distribution and pollen fallout between Bauer Bay and Sandy Bay, Macquarie Island. *Papers and Proceedings of the Royal Society of Tasmania,* **122**(1), 247–53

Peterson, J. A., Scott, J. J. & Derbyshire, E. (1983). Australian landform example No. 43. Sorted stripes of periglacial origin. *Australian Geographer,* **15**(5), 325–8

Phillpot, H. R. (1964). The climate of the Antarctic. In *Antarctic Biology,* ed. R. Carrick, M. W. Holdgate & J. Prévost, pp. 73–9. Paris: Hermann

Pickard, J. & Seppelt, R. D. (1984). Phytogeography of Antarctica. *Journal of Biogeography,* **11,** 83–102

Pittock, A. B. (1971). Rainfall and the general circulation. *Proceedings of the international conference on weather modification.* Canberra: Australian Academy of Science and American Meteorological Society, 330–9

Pollitz, F. F. (1986). Pliocene change in Pacific plate motion. *Nature,* **320,** 738–41

Powell, A. W. B. (1957). Mullosca of Kerguelen and Macquarie Island. *British, Australian and New Zealand Antarctic Research Expedition 1929–1931, Reports, Series B,* **6**(7), 107–50

Powell, A. W. B. (1971). Antarctic mollusca. *Poirieria,* **6**(1), 15–21

Pye, T. (1984). Biology of the house mouse (*Mus musculus*) on Macquarie Island. *Tasmanian Naturalist,* **79,** 6–10

Pye, T. & Bonner, W. N. (1980). Feral brown rats, *Rattus norvegicus,* in South Georgia (South Atlantic Ocean). *Journal of Zoology, (London),* **192,** 237–55

Quilty, P. G., Rubenach, M. & Wilcoxon, J. A. (1973). Miocene ooze from Macquarie Island. *Search,* **4**(5), 163–4

Rand, R. W. (1956). Notes on the Marion Island fur seal. *Proceedings of the Zoological Society of London,* **126,** 65–82

Raper, S. C. B., Wigley, T. M. L., Jones, P. D., Kelly, P. M., Mayes, P. R. & Limbert, D. W. S. (1983). Recent temperature changes in the Arctic and Antarctic. *Nature,* **306,** 458–9

Raven, P. H. & Raven, T. E. (1976). The genus *Epilobium* (Onagraceae) in Australasia: a systematic and evolutionary study. *Department of Scientific and Industrial Research, Bulletin,* **216,** Christchurch: DSIR

Reilly, P. N. & Kerle, J. A. (1981). A study of the gentoo penguin *Pygoscelis papua. Notornis,* **28,** 189–202

Richdale, L. E. (1950). The pre-egg stage in the albatross family. *Biological Monographs,* **3.** Dunedin. 92 pp

Ricker, R. W. (1987). *Taxonomy and Biogeography of Macquarie Island Seaweeds.* London: British Museum (Natural History). 344 pp

Robertson, G. R. (1986). Population size and breeding success of the gentoo penguin *Pygoscelis papua* at Macquarie Island. *Australian Wildlife Research,* **13**(4), 583–7

Rounsevell, D. E. (1983). Return of the Kings. *Aurora,* **9,** 6–8

Rounsevell, D. E. & Copson, G. R. (1982). Growth rate and recovery of a king penguin, *Aptenodytes patagonicus,* population after exploitation. *Australian Wildlife Research,* **9**, 519–25

Salas, M. (1983). Long-distance pollen transport over the southern Tasman Sea: evidence from Macquarie Island. *New Zealand Journal of Botany,* **21**, 285–92

Salinger, M. J. & Gunn, J. M. (1975). Recent climatic warming around New Zealand. *Nature,* **256**, 396–8

Scheffer, V. B. (1958). *Seals, sea lions and walruses. A review of the Pinnipeds.* California: Stanford University Press. 179 pp

Schwerdtfeger, W. (1984). Weather and Climate of the Antarctic. *Developments in Atmospheric Science,* **15**. Amsterdam: Elsevier. 261 pp

Scott, J. H. (1883). Macquarie Island. *Transactions and Proceedings of the New Zealand Institute 1882,* **15**, 484–93

Scott, J. J. (1985). Effects of feral rabbits on the revegetation of disturbed coastal slope sites, Macquarie Island. M.A. thesis, Monash University, Melbourne. 219 pp

Scott, J. J. (1988). Rabbit distribution history and related land disturbance. Macquarie Island. *Papers and Proceedings of the Royal Society of Tasmania,* **122**(1), 255–66

Scott, J. J. (1989). New records of vascular plants from Heard Island. *Polar Record,* **25**(152), 37–42

Segonzac, M. (1972) Données récentes sur la faune des îles Saint-Paul et Nouvelle Amsterdam. *L'oiseau at la revue française d'ornithologie,* **42**, 3–68

Selkirk, D. R., Selkirk, P. M. & Griffin, K. (1983). Palynological evidence for Holocene environmental change and uplift on Wireless Hill, Macquarie Island. *Proceedings of the Linnean Society of New South Wales,* **107**, 1–17

Selkirk, D. R., Selkirk, P. M. & Seppelt, R. D. (1986). An annotated bibliography of Macquarie Island. *ANARE Research Notes,* **38**. 134 pp

Selkirk, D. R., Selkirk, P. M., Bergstrom, D. M. & Adamson, D. A. (1988). Ridge top peats and palaeolake deposits on Macquarie Island. *Papers and Proceedings of the Royal Society of Tasmania,* **122**(1), 83–90

Selkirk, P. M., Adamson, D. A. & Seppelt, R. D. (1988). Terrace form and vegetation dynamics on Macquarie Island. *Papers and Proceedings of the Royal Society of Tasmania,* **122**(1), 59–64

Selkirk, P. M., Costin, A. B., Seppelt, R. D. & Scott, J. J. (1983). Rabbits, vegetation and erosion on Macquarie Island. *Proceedings of the Linnean Society of New South Wales,* **106**(4), 337–46

Selkirk, P. M. & Seppelt, R. D. (1984). Fellfield on Macquarie Island. *Tasmanian Naturalist,* **78**, 24–6

Seppelt, R. D. (1977). Studies on the bryoflora of Macquarie Island I. Introduction and checklist of species. *The Bryologist,* **80**(1), 167–70

Seppelt, R. D. (1980a). A synoptic moss flora of Macquarie Island. *Antarctic Division Technical Memorandum,* **93**. 8pp. Melbourne: Antarctic Division

Seppelt, R. D. (1980b). Bryophytes and lichens collected by the visit of Australian Museum personnel to Macquarie Island, Summer 1977–1978. *Antarctic Division Technical Memorandum,* **94**. 11 pp. Melbourne: Antarctic Division

Seppelt, R. D. (1981). Studies on the bryoflora of Macquarie Island III. Collections, new moss additions and corrections and a revised checklist. *The Bryologist,* **54**(2), 249–52

Seppelt, R. D. & Ashton, D. H. (1978). Studies on the ecology of the vegetation at Mawson Station, Antarctica. *Australian Journal of Ecology*, **3**, 373–88

Seppelt, R. D., Copson, G. R. & Brown, M. J. (1984). Vascular flora of Macquarie Island. *Tasmanian Naturalist*, **78**, 7–12

Shaughnessy, P. & Shaughnessy, G. (1987a). Fur seals at Macquarie Island. *Aurora*, **7**(2), 3–7

Shaughnessy, P. D. (1971). Frequency of the white phase of the southern giant petrel *Macronectes giganteus* (Gmelin). *Australian Journal of Zoology*, **19**, 77–83

Shaughnessy, P. D. (1975). Variation in facial colour of the Royal penguin. *The Emu*, **75**(3), 147–52

Shaughnessy, P. D. & Shaughnessy, G. L. (1987b). Recovery of the fur seal population at Macquarie Island. *1986–87 Australian Antarctic Research Program. Initial Field Reports*, 114–20

Sheard, K. (1953). Taxonomy, distribution and development of the Euphausiacea (Crustacea). *British, Australian and New Zealand Antarctic Research Expedition 1929–1931, Reports, Series B*, **8**(1), 1–72

Simpson, K. G. (1965). The dispersal of regurgitated pumice gizzard-stones by the Southern skua at Macquarie Island. *The Emu*, **65**(2), 119–24

Simpson, R. D. (1972). The ecology and biology of molluscs in the littoral and sub-littoral zones at Macquarie Island, with special reference to *Patinigera macquariensis* (Finlay, 1927). Ph.D. thesis, University of Adelaide, Adelaide. 360 pp

Simpson, R. D. (1976a). Physical and biotic factors limiting the distribution and abundance of littoral molluscs on Macquarie Island (sub-antarctic). *Journal of Experimental Marine Biology and Ecology*, **21**(1), 11–49

Simpson, R. D. (1976b). The shore environment of Macquarie Island. *ANARE Scientific Reports, Series B (1) Zoology*. 41 pp

Simpson, R. D. (1977). The reproduction of some littoral molluscs from Macquarie Island, subantarctic. *Marine Biology (Berlin)*, **44**(2), 125–42.

Simpson, R. D. (1982a). The reproduction of some echinoderms from Macquarie Island. *Australian Museum Memoir*, **16**, 39–52

Simpson, R. D. (1982b). Reproduction and lipids in the sub-antarctic limpet *Nacella* (*Patinigera*) *macquariensis* Finlay, 1927. *Journal of Experimental Marine Biology and Ecology*, **56**(1), 33–48

Simpson, R. D. (1984). The shore ecology of Macquarie Island. *Tasmanian Naturalist*, **78**, 28–30

Simpson, R. D. (1985). Relationship between allometric growth, with respect to shell height, and habitats for two patellid limpets, *Nacella* (*Patinigera*) *macquariensis* Finlay, 1927, and *Cellana tramoserica* (Holten, 1802). *The Veliger*, **28**, 18–27

Simpson, R. D. & Harrington, S. A. (1985). Egg masses of three gastropods, *Kerguelenella lateralis* (Siphonariidae), *Laevilittorina caliginosa* and *Macquariella hamiltoni* (Littorinidae), from Macquarie Island (sub-Antarctic). *Journal of the Malacological Society of Australia*, **7**(1–2), 17–18

Skira, I. J., Brothers, N. P. & Copson, G. R. (1983). Establishment of the European rabbit flea *Spilopsyllus cuniculi* on Macquarie Island, Australia. *Australian Wildlife Research*, **10**(1), 121–7

Smith, R. I. L. (1984). Terrestrial plant biology of the sub-Antarctic and Antarctic. In *Antarctic Ecology*, vol. 1, ed. R. M. Laws, pp. 61–162. London: Academic Press

Smith, V. R. (1976). The effect of burrowing species of Procellariidae on the nutrient status of inland tussock grasslands on Marion Island. *Journal of South African Botany*, **42**, 265–72

Smith, V. R. (1979). The influence of seabird manuring on the phosphorus status of Marion Island (sub-Antarctic) soils. *Oecologia*, **41**, 123–6.

Smith, V. R. (1987a). The environment and biota of Marion Island. *South African Journal of Science*, **83**, 211–20

Smith, V. R. (1987b). Chemical composition of precipitation at Marion Island (sub-antarctic). *Atmospheric Environment*, **21**(5), 1159–65

Sobey, W. R., Adams, K. M., Johnston, G. C., Gould, L. R., Simpson, K. N. G. & Keith, K. (1973). Macquarie Island: the introduction of the European rabbit flea *Spilopsyllus cuniculi* (Dale) as a possible vector for myxomatosis. *Journal of Hygiene, (Cambridge)*, **71**, 299–308

Stonehouse, B. (1956). The Brown Skua *Catharacta skua lonnbergi* (Mathews). *Falkland Islands Dependency Survey, Scientific Reports*, **14**, 1–25

Stonehouse, B. (1960). The King penguin *Aptenodytes patagonica* at South Georgia. 1. Breeding behaviour and development. *Falkland Island Dependencies Survey, Scientific Reports*, **23**, 1–81

Stonehouse, B. (1967). The general biology and thermal balance of penguins. *Advances in Ecological Research*, **4**, 131–96

Stonehouse, B. (1970). Geographic variation in gentoo penguins, *Pygoscelis papua*. *Ibis*. **112**, 52–7

Sykes, L. R. (1967). Mechanism of earthquakes and nature of faulting on the mid-ocean ridges. *Journal of Geophysical Research*, **72**(8), 2133–53

Tattersall, W. M. (1918). Euphausiacea and Mysidacea. *Australasian Antarctic Expedition 1911–1914, Scientific Reports, Series C, Zoology and Botany*, Volume V, Part 5. 15 pp

Taylor, B. W. (1955a). The flora, vegetation and soils of Macquarie Island. *ANARE Reports, Series B, Volume II, Botany*. 192 pp

Taylor, B. W. (1955b). Terrace formation on Macquarie Island. *Journal of Ecology*, **43**, 133–7

Taylor, F. J. (1974a). Chemical analyses of Campbell Island fresh waters. *New Zealand Journal of Marine and Freshwater Research*, **8**(2), 389–462

Taylor, F. J. (1974b). Some freshwater analyses from Auckland Island. *New Zealand Journal of Marine and Freshwater Research*, **8**(4). 709–10

Taylor, R. H. (1979). How the Macquarie Island parakeet became extinct. *New Zealand Journal of Ecology*, **2**, 42–5

Tickell, W. L. N. (1962). The dove prion *Pachyptila desolata* Gmelin. *Falkland Islands Dependencies Survey Scientific Reports*, **33**, 1–55

Tickell, W. L. N. (1968). The biology of the great albatrosses, *Diomedea exulans* and *Diomedea epomorpha*. In *Antarctic Bird Studies*, ed. O. L. Austin (Jr), pp. 1–55. Washington, DC: American Geophysical Union. American Geophysical Union Antarctic Research Series, **12**

Tomkins, R. J. (1985a). Reproduction and mortality of Wandering albatrosses on Macquarie Island. *The Emu*, **85**, 40–2

Tomkins, R. J. (1985b). Attendance of Wandering Albatrosses (*Diomedea exulans*) at a small colony on Macquarie Island. *ANARE Research Notes*, **29**. 20 pp

Tomlin, J. R. Le B. (1948). The Mollusca of Macquarie Island. Gastropods and bivalves. *British, Australian and New Zealand Antarctic Research Expedition 1929–1931, Reports, Series B*, **5**(5), 221–32

Troll, C. (1960). The relationship between the climates, ecology and plant geography of the southern cold temperate zone and of the tropical high mountains. In *A discussion of the biology of the southern cold temperate zone*, (leader) C. F. A. Pantin. *Proceedings of the Royal Society, London, B*, **152**, 529–32

Tyler, P. A. (1972). Reconnaissance limnology of sub-Antarctic islands. I. Chemistry of lake waters from Macquarie Island and the Iles Kerguelen. *International Revue der gesamten Hydrobiologie und Hydrographie*, **57**(5), 759–78

van Aarde, R. J. (1979). Distribution and density of the feral house cat. *Felis catus*, at Marion Island. *South African Journal of Antarctic Research*, **9**, 14–19

van Aarde, R. J. (1980a). Fluctuations in the population of southern elephant seals *Mirounga leonina* at Kerguelen Island. *South African Journal of Zoology*, **15**(2), 99–106

van Aarde, R. J. (1980b). The diet and feeding behaviour of feral cats, *Felis catus*, at Marion Island. *South African Journal of Wildlife Research*, **10**, 123–8

van Aarde, R. J. (1986). A case study of an alien predator (*Felis catus*) on Marion Island: selective advantages. *South African Journal of Antarctic Research*, **16**(3), 113–14

van Aarde, R. J. & Skinner, J. D. (1982). The feral cat population at Marion Island: characteristics, colonization and control. *Colloque sur les Ecosystèmes Subantarctiques, 1981, Paimpont. Comité National Français des Recherches Antarctiques*, **51**, 281–8

van Rensburg, P. J. J. (1986). Control of the Marion Island cat (*Felis catus*) population – why and how. *South African Journal of Antarctic Research*, **16**(3), 110–12

van Rensburg, P. J. J., Skinner, J. D. & van Aarde, R. J. (1987). Effects of feline panleucopaenia on the population characteristics of feral cats on Marion Island. *Journal of Applied Ecology*, **24**, 63–73

Varne, R., Gee, R. D. & Quilty, P. G. J. (1969). Macquarie Island and the cause of oceanic linear magnetic anomalies. *Science*, **166**, 230–2

Varne, R. & Rubenach, M. J. (1972). Geology of Macquarie Island and its relationship to oceanic crust. In *Antarctic Oceanology II. The Australian–New Zealand Sector*, ed. D. E. Hayes, pp. 251–66. Washington DC: American Geophysical Union. American Geophysical Union Antarctic Research Series, **19**

Vestjens, W. J. M. (1963). Remains of the extinct banded rail at Macquarie Island. *The Emu*, **62**, 249–50

Voisin, J. F. (1968). Les Pétrels géants (*Macronectes halli* et *M. giganteus*) de l'île de la Possession. *L'oiseau et la revue français d'ornithologie*, **38**(Special), 95–122

Wace, N. M. (1960). The botany of the southern oceanic islands. In *A discussion of the biology of the southern cold temperate zone* (leader), C. F. A. Pantin. *Proceedings of the Royal Society, London, B*, **152**, 475–90

Waite, E. R. (1916). Fishes. *Australasian Antarctic Expedition 1911–1914, Scientific Reports, Series C, Zoology and Botany*, Volume III, Part 1. 92 pp

Walton, D. W. H. (1986). The biological basis for conservation of subantarctic islands. *Report of the Joint SCAR/IUCN Workshop at Paimpont, France, 12–14 September, 1986*. Rapporteur – D. W. H. Walton. 32 pp

Walton, D. W. H. & Smith, R. I. L. (1973). Status of the alien vascular flora of South Georgia. *British Antarctic Survey, Bulletin,* **36**, 79–97

Warham, J. (1963). The rockhopper penguin, *Eudyptes chrysocome,* at Macquarie Island. *Auk,* **80**, 229–56

Warham, J. (1967). The White-headed petrel, *Pterodroma lessoni,* at Macquarie Island. *The Emu,* **67**(1), 1–22

Warham, J. (1969). Note on some Macquarie Island birds. *Notornis,* **16**(3), 190–7

Warham, J. (1971). Aspects of breeding behaviour in the royal penguin *Eudyptes chrysolophus schlegeli. Notornis,* **18**, 91–115

Washburn, A. L. (1979). *Geocryology.* London: Edward Arnold. 406 pp

Watkins, B. P. & Cooper, J. (1986). Introduction, present status and control of alien species at the Prince Edward Islands, sub-Antarctic. *South African Journal of Antarctic Research,* **16**(3), 86–94

Watson, K. C. (1967). The terrestrial Arthropoda of Macquarie Island. *ANARE Scientific Reports, Series B (1) Zoology.* 90 pp

Westerkov, K. (1960). Birds of Campbell Island. *New Zealand Department of Internal Affairs, Wildlife Publication,* **61**, 1–83

Wickstead, J. H. (1976). Marine Zooplankton. In *Studies in Biology,* **62**. Marine Biological Association of the United Kingdom

Williams, G. R. (1953). The dispersal from New Zealand and Australia of some introduced European Passerines. *Ibis,* **95**, 676–92

Williams, R. (1988). The nearshore fishes of Macquarie Island. *Papers and Proceedings of the Royal Society of Tasmania,* **122**(1), 233–45

Williamson, P. (1978). The palaeomagnetism of outcropping oceanic crust on Macquarie Island. *Journal of the Geological Society of Australia,* **27**(7), 387–94

Williamson, P. & Johnson, B. D. (1974). Crustal structure of the central region of the Macquarie Ridge complex from gravity studies. *Marine Geophysical Researches,* **2**, 127–32

Williamson, P. & Rubenach, M. J. (1972). Preliminary report on geophysical studies on Macquarie Island. In *Antarctic Oceanology II The Australian–New Zealand Sector,* ed. D. Hayes, pp. 243–9. Washington DC: American Geophysical Union. American Geophysical Union Antarctic Research Series, **19**

Williamson, P. E. (1988). Origin, structural and tectonic history of the Macquarie Island region. *Papers and Proceedings of the Royal Society of Tasmania,* **122**(1), 27–43

Wilson, K. L. (1981). A synopsis of the genus *Scirpus* sens. lat. (Cyperaceae) in Australia. *Telopea,* **2**(2), 153–72

Woehler, E. (1984). Morphology of the Royal penguin *Eudyptes schlegeli* at Macquarie Island. *Tasmanian Naturalist,* **79**, 2–4

Zinova, A. D. (1958). The composition and character of algal flora at the Antarctic coast and in the vicinity of Kerguelen and Macquarie Islands. *Soviet Antarctic Expedition Information Bulletin,* **3**, 47–9 (in Russian). (Published in English in *Soviet Antarctic Expedition Information Bulletin,* **1**, 123–5. Amsterdam: Elsevier.)

Zinova, A. D. (1963). Delesseriaceae apud Insulas Kerguelen et Macquarie. *Trudy Botanicheskogo Instituta. Akademii Nauk SSSR, Seriya 2,* **16**, 52–67

APPENDIX 1

Vascular plants recorded from Macquarie Island

After Copson (1984) except where indicated.

Pteridophytes
Blechnum penna-marina (Poir.) Kuhn
Grammitis poeppigeana (Mett.) Pic. Serm.
Hymenophyllum peltatum (Poir.) Desv.
Lycopodium australianum (Herter) Allan
  (1)
Polystichum vestitum (Forst. f.) Presl

Angiosperms, dicotyledons
Acaena magellanica (Lam.) Vahl
Acaena minor (Hoof. f.) Allan
Azorella macquariensis Orchard (2)
Callitriche antarctica Engelm. ex Hegel.
Cardamine corymbosa Hook. f.
Cerastium fontanum Baumg.
Colobanthus muscoides Hook. f.
Colobanthus quitensis Bartl.
Coprosma perpusilla Colenso ssp.
  subantarctica Orchard (3)
Cotula plumosa Hook. f.
Crassula moschata Forst. f.
Epilobium brunnescens (Cockayne) Raven
  et Engelhorn var *brunnescens* (4)
Epilobium pendunculare A. Cunn. (5)
Galium antarcticum Hook. f. (6)
Hydrocotyle sp.
Montia fontana L.
Myriophyllum triphyllum Orchard
Pleurophyllum hookeri Buchan.
Ranunculus biternatus Smith
Rumex crispus L.
Stellaria decipiens Hook. f.
Stellaria media (L.) Vill.
Stilbocarpa polaris (Homb. et Jacq.) Gray

Angiosperms, monocotyledons
Agrostis magellanica Lam.

Anthoxanthum odoratum L. (6)
Carex trifida Cav.
Corybas macranthus (Hook. f.) Reichb. f.
Deschampsia chapmanii Petrie
Deschampsia caespitosa (L.) P. Beauv. (7)
Festuca contracta Kirk
Isolepis aucklandica Hook. f. (8)
Juncus scheuchzerioides Gaudichaud
Luzula crinita Hook. f. var. *crinita* (9)
Poa annua L.
Poa cookii (Hook. f.) Hook. f. (10)
Poa foliosa (Hook. f.) Hook. f.
Poa litorosa Cheeseman (6)
Puccinellia macquariensis (Cheeseman)
  Allan et Jansen
Uncinia divaricata Boott in Hook. f. (6)
  (11)
Uncinia hookeri Boott in Hook. f. (12)

Legend
  (1) *Lycopodium* sp. of Copson
      (1984)
  (2) formerly included in
      Azorella selago (Orchard,
      1988)
  (3) formerly included in
      Coprosma pumila
      (Orchard, 1986)
  (4) *Epilobium nerterioides*
      (Raven and Raven, 1976)
  (5) *Epilobium linnaeoides*
      (Raven and Raven, 1976)
  (6) addition, Seppelt *et al.*
      (1984)
  (7) *Deschampsia penicillata* (E.
      Edgar pers. comm.)

251

(8) *Scirpus aucklandicus*
    (Wilson, 1981)
(9) *Luzula campestris* (Edgar,
    1966)
(10) *Poa hamiltonii* (E. Edgar,
    1986)

(11) *Uncinia meridensis*
    (K. Wilson pers. comm.)
(12) *Uncinia compacta*
    (K. Wilson pers. comm.)
*   species has been described
    as introduced by humans to
    Macquarie Island.

APPENDIX 2

Bryophytes recorded from Macquarie Island

Mosses

After Seppelt (1981) except where indicated.

Achrophyllum dentatum (Hook. f. et Wils.) Vitt et Crosby

Acrocladium chlamydophyllum (Hook. f. et Wils.) C. Muell. et Broth. (1)

Amblystegium serpens (Hedw.) B.S.G.

Andreaea acuminata Mitt.

Andreaea acutifolia Hook. f. et Wils.

Andreaea gainii Card. (1)

Andreaea mutabilis Hook. f. et Wils.

Andreaea nitida Hook. f. et Wils.

Andreaea subulata Harv. in Hook. f.

Aulacomnium palustre (Hedw.) Schwaegr.

Bartramia papillata Hook. f. et Wils.

Blindia robusta Hampe

Blindia seppeltii Bartlett et Vitt (2)

Brachythecium rutabulum (Hedw.) B.S.G.

Brachythecium salebrosum (Web. et Mohr) B.S.G.

Breutelia elongata (Hook. f. et Wils.) Mitt.

Breutelia pendula (Smith) Mitt.

Bryoerythrophyllum recurvirostre (Hedw.) Chen

Bryum amblyodon C. Muell. (1)

Bryum argenteum Hedw.

Bryum billardieri Schwaegr.

Bryum dichotomum Hedw.

Bryum erythrocarpoides C. Muell. et Hampe (1)

Bryum laevigatum Hook. f. et Wils.

Bryum mucronatum Mitt. in Hook. f.

Bryum pseudotriquetrum (Hedw.) Gaertn., Meyer et Scherb. (1)

Calyptrochaeta apiculatus (Hook. f. et Wils.) Vitt

Campylopus clavatus (R. Br.) Hook. f. et Wils.

Campylopus introflexus (Hedw.) Brid.

Ceratodon purpureus (Hedw.) Brid.

Chrysoblastella chilensis (Mont.) Reim. (3)

Conostomum pentastichum (Brid.) Lindb.

Dicranella cardotii (R. Br. ter.) Dix.

Dicranoloma billardieri (Brid.) Par.

Dicranoloma menziesii (Tayl.) Par.

Dicranoloma robustum (Hook. f. et Wils.) Par.

Dicranoweisia antarctica (C. Muell.) Kindb.

Ditrichum brevirostre (R. Br. ter.) Broth.

Ditrichum punctulatum Mitt.

Ditrichum strictum (Hook. f. et Wils.) Hampe

Drepanocladus aduncus (Hedw.) Warnst.

Drepanocladus uncinatus (Hedw.) Warnst.

Entosthodon subattenuatus (Broth.) Par. (4)

Eurhynchium austrinum (Hook. f. et Wils.) Jaeg.

Fissidens rigidulus Hook. f. et Wils.

Funaria hygrometrica Hedw. (1)

Grimmia apocarpa (Hedw.) B.S.G.

Grimmia trichophylla Grev. (1)

Hypnum chrysogaster C. Muell.

Hypnum cupressiforme Hedw.

Isopterygium limatum (Hook. f. et Wils.) Broth.

Lembophyllum divulsum (Hook. f. et Wils.) Lindb. in Par.

253

Leptostomum inclinans R. Br.
Macromitrium longirostre (Hook.)
  Schwaegr.
Muelleriella crassifolia (Hook. f. et Wils)
  Dus.
Mielichhoferia sp.
Orthodontium lineare Schwaegr.
Plagiothecium denticulatum (Hedw.)
  B.S.G.
Philonotis scabrifolia (Hook. f. et Wils.)
  Braithw.
Pohlia sp. 1 (1)
Pohlia sp. 2 (1)
Pohlia wahlenbergii (Web. et Mohr)
  Andrews in Grant
Polytrichum alpinum L. ex Hedw. (5)
Polytrichum juniperinum Willd. ex Hedw.
Pottia heimii (Hedw.) Feurnr.
Psilopilum australe (Hook. f. et Wils.)
  Mitt.
Ptychomnion aciculare (Brid.) Mitt.
Ptychomnion densifolium (Brid.) Jaeg.
Rhacocarpus purpurascens (Brid.) Par.
Rhacomitrium crispulum (Hook. f. et
  Wils.) Hook. f. et Wils.
Rhacomitrium lanuginosum (Hedw.) Brid.
Rhizogonium mnioides (Hook.) Wils.
Sphagnum falcatulum Besch.
Tayloria octoblepharis (Hook. f.) Mitt.
Thuidium furfurosum (Hook. f. et Wils.)
  Reichdt.
Tortella calycina (Schwaegr.) Dix.
Tortula andersonii Aongstr. (6)
Tortula rubra Mitt.
Trematodon flexipes (Hook.) Brid.
Tridontium tasmanicum Hook. f.
Ulota phyllantha Brid.
Verrucidens tortifolius (Hook. f. et Wils.)
  Reim.
Zygodon menziesii (Schwaegr.) Arnott
Note that *Breutelia affinis, Bryum
  amblyolepis, Eurhynchium
  praelongum, Rhynchostegium
  tenuifolium* and *Sematophyllum* sp.
  recorded in Seppelt (1981) are
  incorrect records and have been
  deleted from this list.

Legend
(1) addition

(2) Bartlett and Vitt (1986)
(3) *Cheilothela chilensis*
(4) *Funaria producta* (A. Fife,
    pers. comm.)
(5) *Pogonatum alpinum*
(6) *Tortula* aff. *bealeyensis*
    (Lightowlers, 1986)
* Species has been described as
  introduced by humans to
  Macquarie Island.

Hepatics
After Seppelt (1977) except
  where indicated

Adelanthus occlusus (Hook. f. et Tayl.)
  Carringt.
Adelanthus falcatus (Hook.) Mitt. (1)
Andrewsianthus perigonialis (Hook. f. et
  Tayl.) Schust. (1)
Cephaloziella exiliflora (Tayl.) Dovin (1)
Chandonanthus squarrosus (Hook.) Mitt.
  in Schiffn. (1)
Cheilolejeunea albovirens (Hook. f. et
  Tayl.) Hodgs. (1)
Chiloscyphus sp.
Clasmatocolea humilis (Hook. f. et Tayl.)
  Grolle (1)
Clasmatocolea paucistipula (Rodw.)
  Grolle
Clasmatocolea strongylophylla (Hook. f.
  et Tayl.) Grolle (1)
Clasmatocolea turgescens (Hook. f. et
  Tayl.) Grolle
Clasmatocolea zotovii (Herz.) Grolle
Cryptochila acinacifolia (Hook. f. et
  Tayl.) Grolle
Cryptochila grandiflora (Lindeng. et
  Gott.) Grolle
Fossombronia australis Mitt. (1)
Frullania rostrata (Hook. f. et. Tayl.)
  Hook. f. et Tayl. ex Gottsche,
  Lindenb. et Nees
Gackstroemia weindorferi (Herz.) Grolle
Herzogobryum sp. (1)
Isotachis sp. (1)
Jamesoniella colorata (Lehm.) Spruce
Jungermannia inundata Hook. f. et Tayl.
  (1)

Lepicolea scolopendra (Hook.) Dum. ex
Trev. (1)
Lepidolaena allophylla (Hook. f. et Tayl.)
Trev.
Lepidolaena magellanica (Lam.) Schiffn.
(2)
Lepidozia laevifolia (Hook. f. et Tayl.)
Schust.
Lepidozia sp. (1)
Leptoscyphus australis (Hook. f. et Tayl.)
Schust.
Lophocolea bidentata (L.) Dumort.
Lophocolea bispinosa (Hook. f. et Tayl.)
Gottsche, Lindenb. et Nees
Lophocolea fulva Steph.
Lophocolea lenta (Hook. f. et Tayl.)
Gottsche, Lindenb. et Nees
Lophocolea pallida Mitt. (1)
Marchantia berteroana Lehm. et Lindenb.
Marsupidium perpusillum (Col.) Hodgs.
(1)
Megaceros sp. (1)
Metzgeria atrichoneura Spruce
Metzgeria furcata (L.) Dumort.
Metzgeria hamata Lindb.
Pachyglossa tenacifolia (Hook. f. et Tayl.)
Herz. et Grolle

Paraschistochila sp. (1)
Plagiochila banksiana Gottsche
Plagiochila circinalis (Lehm.) Lehm. et
Lindenb. (3)
Plagiochila ratkowskiana Inoue (3)
Plagiochila retrospectans Nees
Riccardia aequicellularis (Steph.) Hewson
(1)
Riccardia alcicornis (Hook. f. et Tayl.)
Trev. (1)
Riccardia cochleata (Hook. f. et Tayl.)
Kuntze (4)
Riccardia colensoi (Steph.) Martin (1)
Seppeltia succuba Grolle in Grolle et
Seppelt (5)
Temnoma palmatum (Lindb. ex Pears.)
Schust.
Triandrophyllum subtrifidum (Hook. f. et
Tayl.) Fulf. et Hatch.

Legend
(1) addition
(2) doubtful record
(3) Inoue and Seppelt (1985)
(4) *Aneura* sp.
(5) Grolle and Seppelt (1986)

APPENDIX 3

Lichens recorded from Macquarie Island

After Seppelt (1980b) except where indicated

Acarospora sp.
Argopsis megalospora Th. Fr.
Blastenia macquariensis Dodge (2)
Buellia mawsoni Dodge (1)
Caloplaca macquariensis Dodge (2)
Chiodecton acarosporoides Dodge (3)
Chiodecton macquariense Dodge (3)
Cladia aggregata (Sw.) Nyl. (1)
Cladia taylorii Dodge (3)
Cladonia auerii Räsänen
Cladonia cervicornis (Ach.) Flotow (6)
Cladonia chlorophaea (Flörke) Sprengel (6)
Cladonia coniocraea (Flörke) Sprengel
Cladonia corniculata Ahti et Kashiwadani in Inoue (6)
Cladonia cornuta (L.) Hoffm.
Cladonia ecmocyna Leighton (6)
Cladonia furcata (Huds.) Schrader (6)
Cladonia gracilis (L.) Willd. (6)
Cladonia pleurota (Flörke ex Somm.) Schaerer (6)
Cladonia pyxidata (L.) Hoffm. (6)
Cladonia scabriuscula (Delise) Leighton (6)
Cladonia subantarctica Filson et Archer (6)
Cladonia subdigitata Vainio (6)
Coccocarpia kerguelensis Dodge (1)
Coenogonium subtorulosum Müll. Arg. (1)
Fistulariella inflata (Hook. f. et Tayl.) Bowler et Rundel
Gasparrinia macquariensis Dodge (1)
Graphis sp. (8)
Hypogymnia lugubris (Pers.) Krog

Hypotrachyna sinuosa (Sm.) Hale
Kuttlingeria macquariensis Dodge (3)
Lecania johnstoni Dodge (1)
Lecanora brownii Dodge (3)
Lecanora parmelina Zahlbr.
Lecanora prolifera Dodge (2)
Lecidea haysomi Dodge (2)
Lecidea macquariensis Dodge (3)
Lepraria sp. (8)
Mastodia macquariensis Dodge (3)
Mastodia tessellata (Hook. f. et Harv.) Hook. f. et Harv. ex Hook. f.
Menegazzia castanea James et Galloway (7)
Menegazzia subpertusa James et Galloway (7)
Microthelia macquariensis Dodge (1)
Mycoblastus campbellianus (Nyl.) Zahlbr. (1)
Neofuscelia glabrans (Nyl.) Essl. (8)
Omphalodina macquariensis Dodge (3)
Opegrapha macquariensis Dodge (3)
Pannaria sp. (1)
Parmelia brevirhiza Kurokawa (4)
Parmelia brownii Dodge (3)
Parmelia cunninghamii Crombie
Parmelia haysomii Dodge (4)
Parmelia labrosa (Zahlbr.) Hale (4)
Parmelia lusitaniensis Filson (4)
Parmelia macquariensis Dodge (4)
Parmelia phillipsiana Filson (4)
Parmelia signifera Nyl. (4)
Parmelia sp.
Parmelia sulcata Tayl. (4)
Parmelia texana Tuck. (4)
Parmelia waiporiensis Hillm. (4)

256

Peltigera horizontalis (Huds.) Baumg.
Peltigera rufescens (Weis) Humb.
Peltigera sp.
Peltularia crassa Jørg. et Galloway (7)
Pertusaria dactylina (Ach.) Nyl.
Pertusaria tylopaca Nyl.(1)
Phyllopyrenia macquariensis Dodge (2)
Phlyctis macquariensis Dodge (2)
Physcia macquariensis Dodge (2)
Physcia adscendens (Fr.) Oliv. (3)
Porina macquariensis Dodge (3)
Pseudocyphellaria delisea (Delise)
    Galloway et James
Pseudoparmelia caperata (L.) Hale
Psoroma hypnorum (Vahl) S. F. Gray
Psoroma macquariensis Dodge (3)
Psoroma versicolor Müll. Arg. (1)
Pyrenodesmia inclinans (Stirton) Dodge
    (1)
Pyrenodesmia subpyracea (Nyl.) Dodge
    (1)
Ramalina banzarensis Dodge (1)
Ramalina farinacea Des Abbayes (1)
Ramalina inflata (Hook. f. et Tayl.) Hook.
    f. et Tayl. (1)
Ramalina unilateralis F. Wilson (1)
Rinodina peloleuca (Nyl.) Müll. Arg. (1)
Rinodina subbadioatra (Knight) Dodge (1)
Siphulastrum cladinoides Dodge (1)
Siphulastrum mamillatum (Hook. f. et
    Tayl.) Galloway
Siphulastrum usneoides Dodge (1)
Sphaerophorus curtus Hook. f. et Tayl. (3)
Sphaerophorus globosus (Huds.) Vain
Sphaerophorus melanocarpus (Sw.) DC.
Sphaerophorus ramulifer Lamb
Sphaerophorus sp.

Sphaerophorus tener Laurer (5)
Sporastia desmaspora (Knight) Dodge (3)
Squamarina haysomii Dodge (3)
Steinera neozelandica Dodge (3)
Stereocaulon argus Hook. f. et Tayl.
Stereocaulon corticulatum Nyl.
Stereocaulon leptaleum Nyl. (1)
Stereocaulon macquariense Dodge (2)
Stereocaulon pulvinare Dodge (1)
Stereocaulon ramulosum (Sw.) Räuschel
Stereocaulon submollescens Nyl. (1)
Sticta martinii Galloway (7)
Thamnolia vermicularis (Sw.) Schaerer
Thelidea sp. (1)
Toninia sp.
Usnea antarctica Du Reitz (1)
Usnea arida Mot. (8)
Usnea contexta Mot.
Usnea glomerata Mot.
Usnea laxissima Dodge (1)
Usnea torulosa (Müll. Arg.) Zahlbr. (8)
Usnea xanthopoga Nyl. (8)
Usnea sp. 1
Usnea sp. 2
Verrucaria durietzii Lamb (1)
Xanthoria elegans (Link.) Th. Fr.

Legend
(1) from Dodge (1948)
(2) from Dodge (1968)
(3) from Dodge (1970)
(4) from Filson (1981)
(5) from Filson (1986)
(6) from Filson and Archer
    (1986)
(7) D. Galloway det.
(8) addition

APPENDIX 4

Fungi recorded from Macquarie Island

After Kerry (1984)

Basidiomycetes
?*Camarophyllus* sp.
Cantharellus sp.
Clavaria sp.
Clitocybe sp.
Coprinus disseminatus (Pers. ex Fr.) S. F.
    Gray
Cystoderma amianthinum (Scop. ex Fr.)
    Fayod ex aut.
Cystoderma carcharis (Pers. ex Secr.)
    Fayod
Eccilia sp.
Galera sp.
Galerina longinqua Smith et Singer
Galerina macquariensis Smith et Singer
Galerina vittiformis (Fr.) Singer
Galerina spp.
Gerronema schusteri Singer
Hygrocybe conica (Scop. ex Fr.) Kummer
    et Führ
Hygrocybe nigrescens (Quél.) Kühn.
Hygrocybe aff. *ceracea* (Wulf. ex Fr.)
    Kummer
Hygrocybe spp.
Hygrophorus conicus Scop. ex Fr.
Lepista fibrosissima Singer
Mycena metata (Fr. ex Fr.) Kummer
Mycena sp.
Naucoria sp.
Nolanea spp.
Omphalina sp.
Panaeolus moellerianus Singer
Panaeolus papilionaceus (Bull. ex Fr.)
    Quél.
Panaeolus sp.
Phaeogalera stagnina (Fr.) Pegler et
    Young
Pholiota myosotis (Fr. ex Fr.) Singer
Pholiota sp.
Psathyrella macquariensis Singer
Psathyrella sp.
Psilocybe longinqua Singer
Puccinia sp.
Rhodophyllus sericeus (Bull. ex Mérat.)
    Quél. ssp. *antarcticus* (Singer) Singer
?*Tricholoma* sp.
Uredo sp.

Ascomycetes
Anthostomella phaeosticta (Berk.) Sacc.
    (1)
Ceratosphaeria sp.
Chaetomium sp.
Dasyscyphus cf. *enzenspergerianus* (P.
    Henn.) Dennis
Didymella proximella (Karst.) Sacc.
Dothidea spilomea Berk.
?*Fabraea* sp.
Hendersonia microsticta Berk.
?*Hymenoscyphus* sp.
Hypodermella sp.
Leptosphaeria eustoma (Fuckel) Sacc.
Leptosphaeria doliolum (Pers. ex Fr.) Ces.
    et De Not.
Leptosphaeria sp.
Mycosphaerella tassiana (De Not.) Joh.
Mycosphaerella sp. 1
Mycosphaerella spp.
Lachnea sp.
Niesslia exosporioides (Desm.) Wint.
Pleospora graminearum Wehm.
Pleospora heleocharidis Karst.

Pleospora herbarum (Fr. ex Fr.) Rabenh.
  (2)
Pleospora lutea Wehm.
Pleospora vagans Neissl.
Pleospora sp.
Sarcoscypha sp.
Schizothyrioma stilbocarpae Spooner
Sphaeria depressa Sow.

Imperfect fungi (Coelomycetes)

Ascochyta stilbocarpae Syd.
Camarosporium metableticum Trial.
Camarosporium sp.
Colletitrichum gloeosporioides Penz.
Diploidia sp.
Phoma exigua Desm.
Phoma ?herbarum Westend
Phoma sp. 1
Phoma sp. 2
Phoma sp. 3
Phyllosticta sp.
Rhodesiopsis gelatinosa gen. et sp. nov.
  ined.
Septoria sp.
Stagonospora ischmaemi Sacc.
Tunicago sp.

Imperfect fungi (Hyphomycetes)

Acremonium kiliense Grütz.
Acremonium strictum W. Gams
Acremonium terricola (Miller *et al.*) W.
  Gams
Acremonium ?zonatum (Sawada) W.
  Gams
Acremonium spp.
Alternaria alternata (Fr.) Keissler
Aspergillus sydowi (Bayner et Sartory)
  Thom et Church
Aureobasidium pullulans (de Bary)
  Arnaud
Botrytis cinerea (Pers. ex Fr.) Rabenh.
Cephalosporium sp.
Chalara sp.
Chrysosporium pannorum (Link) Hughes

Cladosporium cladosporoides (Fresen.) de
  Vries
Cladosporium herbarum (Pers.) Link
Cladosporium spp.
Curvularia trifolii (Kauffm.) Boedjn
Cylindrophora sp.
?Diploospora sp.
Epicoccum purpurascens Ehrenb. ex
  Schlecht.
Fusidium sp.
Gliocladium sp.
Graphium sp.
Hyalopus sp.
Monosporium sp.
Paecilomyces sp.
Penicillium brevi-compactum Dierckx
Penicillium corylophilum Dierckx
Penicillium cyclopium Westling
Penicillium decumbens Thom.
Penicillium frequentans Westling
Penicillium granulatum Bain
Penicillium restrictum Gilman et Abbot
Penicillium spp.
Polyscytalum sp.
Spicaria sp.
Sporotrichum sp.
Stemphyllium botryosum Wallr.
Trichoderma harzianum Rifae
Trichoderma sp.
Verticillium lecanii (Zimm.) Viégas
Volucrispora graminea Ingold, McDougall
  et Dann

Lower fungi (Zygomycetes)

Mortierella gamsii Mil'ko
Mortierella spp.
Mucor hiemalis Wehmer f. *hiemalis*
Mucor mucedo (L.) Fresenius
Mucor sp.
Rhizopus nigricans Ehrenb.

Legend
(1) Sphaeria phaeosticta (Berk.)
(2) Sphaeria herbarum Fr. ex Fr.

APPENDIX 5

Freshwater and terrestrial algae recorded from Macquarie Island

Cyanophyta (blue-green algae)
Nostoc sp. (2)
Nodularia sp. (2)

Chlorophyta (green algae)
Cosmarium 3 spp. (3)
Closterium 5 spp. (3)
Cylindrocystis sp. (3)
Euastrum sp.(3)
Netium sp. (3)
Oedogonium sp. (3)
Pediastrum sp. (3)
Pleurotaenium sp. (3)
Rhizoclonium sp. (2)
Sphaerozosma sp. (3)
Spirogyra sp. (2)
Staurastrum sp. (2)
Staurodesmus sp. (3)
Stigeoclonium sp. (3)
Tetmemorus sp. (3)
Tribonema sp. (3)
Zygnema sp. (3)

Bacillariophyta (diatoms)
Achnanthes affinis Grun. (1)
Achnanthes biasolettiana (Kütz.) Grun. (1) (2)
Achnanthes brevipes Agardh var. *intermedia* (Kütz.) Cleve (1) (2)
Achnanthes exigua Grun. (1)
Achnanthes exilis Kütz. (1)
Achnanthes lanceolata (Bréb.) Grun. (1)
Achnanthes linearis Grun. (1)
Achnanthes minutissima Kütz. var. *cryptocephala* Grun. (1)
Achnanthes subsalsa Petersen (1)
Amphora delicatissima Krasske (1) (2)
Amphora ovalis Kütz. (2)
Cocconeis placentula Ehrenb. (1) (2)
Coscinodiscus spp. (1) (fragments)
Cyclotella antiqua W. Smith (2)
Cyclotella meneghiniana Kütz. (2)
Cyclotella stelligera Cleve et Grunow (2)
Cymatopleura solea (Bréb.) W. Smith (2)
Cymatopleura solea var. *regula* (Ehrenb.) Grun. (1)
Cymbella americana A. Schm. (2)
Cymbella pusilla Grun. (2)
Cymbella pusilla var.? (1)
Diatomella sp. (2)
Diatomella balfouriana (Grev.) Agardh (1) (2)
Diploneis smithii (Bréb.) Cleve (1) (2)
Encyonema gracile Rab. (2)
Encyonema lunula Grun. (2)
Epithemia sorex Kütz. (2)
Eunotia alpina (Naeg.) Hustedt (1) (2)
Eunotia arcus Ehrenb. (2)
Eunotia lunaris (Ehrenb.) Grun. (1) (2)
Eunotia monodon Ehrenb. (2)
Eunotia pectinalis (Kütz.) Rabenh. (2)
Eunotia robusta Ralfs (2)
Eunotia trinacria Krasske (1) (2)
Eunotia triodon Ehrenb. (2)
Eunotia zygodon Ehrenb. (2)
Fragilaria bicapitata A. Mayer (1) (2)
Fragilaria capucina Desmaz. (1) (2)
Fragilaria harrisonii (W. Sm.) Grun. (1) (2)
Fragilaria pinnata Ehrenb. (1) (2)
Fragilaria virescens Ralfs (1) (2)

Fragilariopsis antarctica (Castr.) Hustedt (1)

Frustulia minuta Rabenh. (2)

Frustulia rhomboides (Ehrenb.) De Toni (1) (2)

Gomphonema angustatum (Kütz.) Rabenh. var. *productum* Grun.

Gomphonema intricatum Kütz. (1) (2)

Gomphonema parvulum (Kütz.) Grun. (1) (2)

Hantzschia amphioxys (Ehrenb.) Grun. (1) (2)

Mastogloia sp. (2)

Mastogloia ?squamosa Brun. (2)

Melosira crenulata Kütz. (2)

Melosira decipiens Grove (2)

Melosira granulata (Ehrenb.) Ralfs. (1)

?Melosira sp. (1)

Navicula dicephala Ehrenb. (1)

Navicula mutica Kütz. (1)

Navicula mutica var. *cohnii* (Hilse) Grun. (1)

Navicula mutica var. *nivalis* (Ehrenb.) Hustedt (1)

Navicula mutica var. *truncata* Peragallo (1)

Navicula radiosa Kütz. (2)

Navicula reinhardti Grun. (2)

Navicula ?rostellata Kütz. (1) (2)

Navicula seminulum Grun. (1)

Neidium affine (Ehrenb.) Cleve (1) (2)

Opephora martyi Hériband (1) (2)

Pinnularia appendiculata (Agardh) Cleve (1) (2)

Pinnularia atwoodii Per. (1) (2)

Pinnularia borealis Ehrenb. (1) (2)

Pinnularia brevicostata Cleve (1) (2)

Pinnularia cardinalis (Ehrenb.) W. Smith (1) (2)

Pinnularia commutata Grun. (2)

Pinnularia divergens W. Smith (1) (2)

Pinnularia divergentissima (Grun.) Cleve (1) (2)

Pinnularia fasciata Lagerst. (1) (2)

Pinnularia interrupta W. Smith (1)

Pinnularia lata (Bréb.) W. Smith (1) (2)

Pinnularia macilenta Ehrenb. (2)

Pinnularia maior (Kütz.) Cleve (2)

Pinnularia microstauron (Ehrenb.) Cleve (1) (2)

Pinnularia molaris Grun. (1) (2)

Pinnularia nivorum Per. (1) (2)

Pinnularia rupestris Hantzsch. (2)

Pinnularia viridis (Nitzsch) Ehrenb. (2)

Rhopalodia gibba (Ehrenb.) O. Müller (2)

Rhopalodia gibberula (Ehrenb.) O. Müller (1) (2)

Rhopalodia ventricosa O. Müller (2)

Stauroneis acuta W. Smith (1) (2)

Stauroneis anceps Ehrenb. (1)

Stauroneis anceps var. *hyalina* Brun et Per. (2)

Stauroneis fulmen Brit. (2)

Stauroneis parvula (Grun.) Jan (1)

Stauroneis phoenicenteron Ehrenb. (2)

Stauroneis pygmaea Krieger (1) (2)

Stenopterobia anceps Bréb. (2)

Surirella angustior O. Müller (2)

Surirella angustata Kütz. (1) (2)

Surirellu bifrons Kütz. (1) (2)

Surirella engleri O. Müller (2)

?Surirella engleri forma *angustior* O. Müller (1)

Synedra sp. (2)

Synedra vaucheriae Kütz. (1) (2)

Tabellaria flocculosa (Roth) Kütz. (2)

Legend

(1) Bunt (1954b)

(2) Evans (1970)

(3) Croome (1984)

APPENDIX 6

Marine algae recorded from Macquarie Island

After Ricker (1987)

Green algae (Chlorophyta)

Acrosiphonia pacifica (Montagne) J. Agardh in Hohenacker
Blidingia minima (Nageli ex Kützing) Kylin
Codium subantarcticum Silva sp. ined.
Derbesia furcata Ricker
Endophyton atroviridis O'Kelly sp. ined.
Enteromorpha bulbosa (Suhr) Montagne
Enteromorpha intestinalis (L.) Nees
Entocladia maculans (Cotton) Papenfuss
Monostroma sp. 1
Monostroma sp. 2
Prasiola crispa (Lightfoot) Meneghini
Rhizoclonium ambiguum (Hook. f. et Harv.) Kützing
Ulva angusta Setchell et Gardner
Ulva rigida C. Agardh
Urospora penicilliformis (Roth) Areschoug

Brown algae (Phaeophyta)

Adenocystis utricularis (Bory de Saint-Vincent) Skottsberg
Alethocladus corymbosus (Dickie) Sauvageau
Asteronema ferruginea (Harvey) Delepine et Asensi
Chordariopsis capensis (C. Agardh) Kylin
Cladothele striarioides (Skottsberg) Zinova
Desmarestia chordalis Hook. f. et Harv.
Desmarestia willii Reinsch
Dilophus decumbens Ricker
Durvillaea antarctica (Chamisso) Hariot
Ectocarpus constanciae Hariot
Ectocarpus siliculosus (Dillwyn) Lyngbe

Elachista antarctica Skottsberg
Germinocarpus geminatus (Hook. f. et Harv.) Skottsberg
Gononema pectinatum (Skottsberg) Kuckuck et Skottsberg in Skottsberg
Gononema ramosum (Skottsberg) Kuckuck et Skottsberg in Skottsberg
Haplogloia moniliformis Ricker
Lithoderma piliferum Skottsberg var. *debile* Ricker
Macrocystis pyrifera (L.) C. Agardh
Microzonia velutina (Harvey) J. Agardh
Myrionema incommodum Skottsberg
Papenfussiella tristanensis Kylin
Petalonia fascia (O. F. Müller) Kuntze
Pilayella littoralis (L.) Kjellman
Ralfsia australis Skottsberg
Scytosiphon lomentaria (Lyngbe) Link
Scytothamnus fasciculatus (Hook. f. et Harv.) Cotton
Sphacelaria bornetii Hariot
Streblonema codiatile Ricker

Red algae (Rhodophyta)

Acrochaetium daviesii (Dillwyn) Nägeli
Ahnfeltia plicata (Hudson) Fries
Antithamnionella alternans Ricker
Antithamnionella ternifolia (Hook. f. et Harv.) Lyle
Ballia callitricha (C. Agardh) Kützing
Bostrychia vaga Hook. f. et Harv.
Callithamnion gracile Hook. f. et Harv. in Harv. et Hook. f.
Callocolax neglectus Schmitz ex Batters
Callophyllis variegata (Bory de Saint-Vincent) Kützing

Cenacrum subsutum Ricker et Kraft
Ceramium strictum (Kützing) Harvey
Chaetangium fastigiatum (Bory de
  Saint-Vincent) J. Agardh
Cladodonta lyallii (Hook. f. et Harv.)
  Skottsberg
Colacodasya inconspicua (Reinsch)
  Schmitz in Schmitz et Falkenberg
Dasyptilon pellucidum (Harvey) G.
  Feldmann
Dasyptilon sp.
Delesseria lancifolia (Hook. f.) J. Agardh
Delisea pulchra (Greville) J. Agardh
Heterosiphonia berkeleyi Montagne
Hildenbrandia lecannellieri Hariot
Hymenena decumbens Levring
Hymenena laciniata (Hook. f. et Harv.)
  Kylin
Iridaea cordata (Turner) Bory de
  Saint-Vincent
Iridaea remuliformis Ricker
Lophurella hookeriana (J. Agardh)
  Falkenberg
Medeiothamnion flaccidum (Hook. f. et
  Harv.) Brauner
Melobesia membranacea (Esper)
  Lamouroux
Mesophyllum patena (Hook. f. et Harv.)
  Ricker
Mesophyllum schmitzii (Hariot) Mendoza
Myriogramme livida (Hook. f. et Harv.)
  Kylin
Pachymenia cf. *stipitata* J. Agardh
Palmaria decipiens (Reinsch) Ricker
Palmaria georgica (Reinsch) Ricker
Phyllophora appendiculata Skottsberg in
  Kylin et Skottsberg

Plocamium hookeri Harvey in Hook. f. et
  Harv.
Plocamium secundatum (Kützing) Kützing
Plumariopsis eatoni (Dickie) De Toni
Polycoryne radiata Skottsberg in Kylin et
  Skottsberg
Polysiphonia anisogona Hook. f. et Harv.
Porphyra columbina Montagne
Porphyra ionae Ricker
Porphyra plocamiestris Ricker
Porphyra woolhousiae Harvey
Porphyra sp.
Porphyropsis coccinea (J. Agardh ex J. E.
  Areschoug) Rosenvinge
Pseudolithophyllum consociatum (Foslie)
  Lemoine
Pseudolithophyllum discoideum (Foslie)
  Lemoine
Pseudophycodrys phyllophora (J. Agardh)
  Skottsberg
Pterothamnion plumula (Ellis) Nägeli in
  Nägeli et Cramer
Pterothamnion simile (Hook. f. et Harv.)
  Nägeli
Ptilonia willana Lindauer
Pugetia delicatissima Norris
Rhodochorton concrescens Drew
Rhodochorton purpureum (Lightfoot)
  Rosenvinge
Rhodochorton variabile Drew
Rhodophyllis acanthocarpa (Harvey) J.
  Agardh
Rhodymenia subantarctica Ricker
Schizoseris condensata (Reinsch) Ricker
Schizoseris dichotoma (Hook. f. et Harv.)
  Kylin
Sporoglossum lophurellae Kylin in Kylin
  et Skottsberg

APPENDIX 7

Terrestrial invertebrates (excluding parasites and mites) of Macquarie Island

by Penelope Greenslade, CSIRO, Division of Entomology, GPO Box 1700, Canberra, ACT 2601, Australia

For a detailed treatment of mites and parasitic arthropods on Macquarie Island refer to Watson (1967).

Status of species on Macquarie Island
(new record) – new record for island
(endemic) – endemic species
(status unknown) – status of species on island unknown
(indigenous) – indigenous
(naturalised) – recently introduced, naturalised
(not naturalised) – introduced, not naturalised

Oligochaeta
Megascolecidae: Earthworms
Microscolex macquariensis (Beddard, 1896) (endemic)

Lumbricidae
Bimastus tenuis (Eisen, 1874)

Enchytraeidae
Enchytraeus albidus Henle, 1837 (indigenous)
Lumbricillus macquariensis (Benham, 1905) (indigenous)
Lumbricillus maximus Michaelsen, 1900 (indigenous)
Lumbricillus werthi (Michaelsen, 1905)
Marionina antipodum (Benham, 1905)

Tubificidae
Macquaridrilus bennettae Jamieson, 1968 (endemic)

Mollusca
Athoracophoridae
Pseudaneitea huttoni (Suter, 1909) (indigenous)
Reflectopallium martensi (Suter, 1909) (indigenous)

Limacidae: Slugs
Deroceras reticularis Mueller, 1774 (naturalised)

Punctidae: Snails
Phrixgnathus hamiltoni Suter, 1896 (indigenous)

Arthropoda
Collembola: Springtails
Hypogastruridae
Hypogastrura (Ceratophysella) denticulata (Bagnall, 1941) (naturalised)
Hypogastrura (Hypogastrura) purpurescens (Lubbock, 1868) (naturalised)
Hypogastrura (Hypogastrura) viatica (Tullberg, 1872) (?naturalised)

Neanuridae
Friesea tilbrooki Wise, 1970 (indigenous)
Friesea sp. (status unknown)

Onychiuridae
Mesaphorura sp. *krausbaueri* group Börner, 1901 (not naturalised)
Onychiurus sp. (not naturalised)
Tullbergia bisetosa Börner, 1902 (indigenous)

Tullbergia templei Wise, 1967 (indigenous)

Isotomidae
Archisotoma sp. (indigenous, new record)
Cryptopygus antarcticus antarcticus
  Willem, 1901 (indigenous)
Cryptopygus dubius Deharveng, 1981
  (indigenous)
Cryptopygus caecus Wahlgren, 1900
  (indigenous)
Cryptopygus lawrencei Deharveng, 1981
  (indigenous)
Cryptopygus tricuspis Enderlein, 1909
  (indigenous)
Isotoma (*Parisotoma*) *insularis*
  Deharveng, 1981 (indigenous)
Isotoma (*Folsomotoma*) *punctata*
  Wahlgren, 1902 (indigenous)
Isotoma (*Desoria*) *tigrina* Nicolet, 1842
  (naturalised)
Isotoma (*Pseudosorensia*) sp. (status
  unknown)
Isotoma sp. (status unknown)
Proisotoma sp. (status unknown)

Entomobryidae
Lepidobrya mawsoni (Tillyard, 1920)
  (indigenous)
Lepidocyrtus sp. (status unknown)

Sminthuridae
Katianna banzare Salmon, 1964
  (indigenous)
Polykatianna davidi (Tillyard, 1920)
  (indigenous)
Sminthurides sp. group (not naturalised)
Sminthurinus kerguelenensis Salmon, 1964
  (indigenous)
Sminthurinus quadrimaculatus (Ryder,
  1879) (not naturalised)
Sminthurinus spp. (status unknown, new
  record)
Neelidae
Megalothorax sp. or spp. (?indigenous,
  status unknown)

Insecta
Psocoptera: Booklice
Philotarsidae
Austropsocus insularis Smithers, 1962
  (indigenous)

Thysanoptera: Thrips
Thripidae
Physemothrips chrysodermus Stannard,
  1962 (indigenous)

Hemiptera
Aphididae: Aphids
Jacksonia papillata Theobald, 1923
  (naturalised)
Mysus sp. (status unknown)
Rhopalosiphum padi (Linnaeus, 1758)
  (naturalised)

Diptera: Flies
Sciaridae: Black fungus gnats
Bradysia watsoni Colless, 1962 (status
  unknown)

Tethinidae: Kelp flies
Amalopteryx maritima Eaton, 1875
  (indigenous)
Apetaenus watsoni Hardy, 1962 (endemic)

Coelopidae: Kelp flies
Coelopella curvipes (Hutton, 1902)
  (indigenous)
Coelopella debilis Lamb, 1909
  (indigenous)

Ephydridae
Ephydrella macquariensis (Womersley,
  1937) (endemic)

Carnidae
Australimyza macquariensis (Womersley,
  1937) (endemic)

Dolichopodidae
Schoenophilus pedestris pedestris Lamb,
  1909 (endemic)

Helicomyzidae
Paractora asymmetrica Enderlein, 1930
  (not naturalised)

Chloropidae
Thyridula sp. (status unknown)

Tipulidae: Crane flies
Erioptera (*Trimicra*) *pilipes macquariensis*
  Alexander, 1962 (indigenous)

Chironomidae
Smittia sp. (status unknown)
Telmatogeton macquariensis (Brundin, 1962) (endemic)

Psychodidae: Moth flies
Psychoda alternata Say, 1829 (status unknown)
Psychoda parthenogenetica (Tonnoir, 1940) (?naturalised, status unknown)
Psychoda surcoufi Tonnoir, 1922 (indigenous, status unknown)

Hymenoptera
Diapriidae
Antarctopria latigaster Brues, 1920 (indigenous) Wingless wasp

Scelionidae
1 species collected

Coleoptera
Staphylinidae: Rove beetles
Omaliinae
Crymus sp. (status unknown)
Homalium variipenne Tillyard, 1920 (status unknown)
Omaliomimus albipenne (Kiesenwetter, 1877) (indigenous)
Omaliomimus venator Broun, 1919 (indigenous)
Stenomalium helmsi (Cameron, 1945) (indigenous)
Stenomalium sulcithorax (Broun, 1880) (indigenous)

Aleocharinae
Halmaeusa antarctica Kiesenwetter, 1877 (indigenous)

Byrrhidae: Byrrid beetles
Epichorius sorenseni (Brookes, 1951) (indigenous)

Lepidoptera
Pyralidae: Pyralid moth
Eudonia mawsoni (Womersley & Tindale, 1937) (indigenous)

Arachnida
Araneae
Desidae
Myro kerguelenensis Cambridge, 1876 (indigenous)

Lynyphiidae
Haplinus rufocephalia (Urquart, 1888) (indigenous)
Parafroneta marrineri (Hogg, 1909) (indigenous)

Hahniidae
Hahnia sp. (status unknown)

Tardigrada
Macrobiotidae
Pseudobiotus (*Isohypsibus*) *augusti* (John Murray, 1907) (indigenous)
A total of forty or more species of tardigrades has been collected.

APPENDIX 8

Marine and littoral invertebrates recorded from Macquarie Island and the Macquarie Ridge

After Dawson (1988) except where indicated

Stylasterina
Allopora eguchii Boschma, 1966
Calyptopora reticulata Boschma, 1968
Conopora pauciseptata Broch, 1951
Crypthelia fragilis Cairns, 1983
Errina (Errina) cheilopora Cairns, 1983
E. (Errina) gracilis von Marenzeller, 1903
E. (Inferiolabiata) fascicularis Cairns, 1983
Lepidopora sarmentosa Boschma, 1968

Scleractinia
Aulacyathus recidivus (Dennant, 1906)
Desmophyllum cristagalli Milne Edwards & Haime, 1848
Enallopsammia sp. cf. *E. marenzelleri* Zibrowius, 1973
E. rostrata (Pourtales, 1878)
Flabellum apertum Moseley, 1876
F. knoxi Ralph & Squires, 1962
Fungiacyathus fragilis Sars, 1872
Lophelia prolifera (Pallas, 1766)
Madrepora oculata Linnaeus, 1758
Solenosmilia variabilis Duncan, 1873
Stenocyathus vermiformis (Pourtales, 1868)

Mollusca
Bivalvia
Acesta sp.*
Chlamys patagonica deliculata (Hutton, 1873)
Gaimardia trapesina coccinea Hedley, 1916
Hiatella arctica (Linnaeus, 1767)
Kidderia bicolor (Martens, 1835)
K. oblonga (Smith, 1899)

K. subquadrata (Pelseneer, 1903)
Lasaea rubra rossiana Finlay, 1928
Limatula pygmaea (Philippi, 1845)
Limopsis sp.*
Modiolus areolatus (Gould, 1850)
Monia sp.*
Mysella charcoti (Lamy, 1906)
M. macquariensis (Hedley, 1916)
Neolepton powelli Dell, 1964
Philobrya hamiltoni (Hedley, 1916)
Pleuromeris sp.
Pronucula mesembrina (Hedley, 1916)
Tawera mawsoni (Hedley, 1916)
Thracia sp.*

Gastropoda (excluding pteropods)
Actinoleuca campbelli macquariensis (Hedley, 1916)
Admete harpovoluta (Powell, 1957)
Archidoris kerguelensis (Berg, 1884)
Brookula sp.*
Calliostoma n. sp. (Marshall MS)
Cantharidus (Plumbelenchus) capillaceus coruscans (Philippi, 1848)
Cantrainea inexpectata (Marshall, 1979)
Cymatona kampyla tomlini (Powell, 1955)
C. kampyla kampyla (Watson, 1885)*
Eatoniella sp. cf. *E. kerguelenensis kerguelenensis* (Smith, 1875)
Emarginula sp.
Eumetula macquariensis (Tomlin, 1948)
Evalea sp.
Falsilunatia pisum (Hedley, 1916)
F. sp.

267

Fusitron cancellatus laudandus (Finlay, 1927)
Kerguelenella lateralis (Gould, 1846)
Laevilittorina caliginosa (Gould, 1849)
Lamellaria sp. cf. *L. conica* (Smith, 1902)
Macquariella hamiltoni (Smith, 1898)
Nacella (*Patinigera*) *macquariensis* Finlay, 1927 (2)
N. concinna (Strebel, 1908) (2)
Notoacmea pileopsis sturnus (Hombron & Jacquinot, 1841)
Ovirissoa ainsworthi (Hedley, 1966)
O. nivosa Powell, 1957
Pareuthria sp. cf. *P. valdiviae* (Thiele, 1925)
Prosipho, sp. cf. *P. certus* Thiele, 1912
P. tomlini Powell, 1957
Puncturella pseudanaloga Powell, 1957
Sinezona subantarctica (Hedley, 1916)
Socienna sp.*
Subonoba sp.
Trinchesia macquariensis Burn, 1973
Trophon macquariensis Powell, 1957
T. mawsoni Powell, 1957
Turbonilla (Chemnitzia) lamyi Hedley, 1916
Xanthodaphne sp.*

Amphineura
Hemiarthrum setulosum (Dall, 1876)
Ischnochiton mawsoni Cotton, 1937
Plaxiphora aurata campbelli Filhol, 1880
Terenochiton fairchildi Iredale & Hull, 1929

Brachiopoda
Gyrothyris mawsoni mawsoni Thompson, 1918
G. mawsoni antipodensis Foster,1974
Magellania macquariensis Thompson, 1918
Neocrania sp.

Echinodermata
Comatulides dawsoni McKnight, 1977
Daidalometra arachnoides (A. H. Clark,1909)
Florometra austini A. M. Clark, 1967
Glyptometra inaequalis (Carpenter, 1888)
Metacrinus wyvilli Carpenter, 1888

Ptilocrinus sp. McKnight, 1984
Antedonid Speel & Dearborn, 1983

Asteroidea
Anasterias directa (Koehler, 1920)
A. mawsoni (Koehler, 1920)
A. sphoerulata (Koehler, 1920)
Ceramaster lennoxkingi McKnight, 1973
Cycethra macquariensis Koehler, 1920
Henricia aucklandiae Koehler, 1920
H. lukinskii (Farquhar, 1895)
H. obesa (Sladen, 1889)
Odontaster aucklandensis McKnight,1973
Porania antarctica antarctica Smith, 1876
Pseudarchaster sp. McKnight, 1984
Pteraster (*Apterodon*) *stellifer stellifer* Sladen, 1889
Sclerasterias mollis Hutton, 1872
Smilasterias irregularis H. L. Clark, 1928

Echinoidea
Goniocidaris umbraculum Hutton, 1878
Pseudechinus novaezealandiae (Mortensen, 1921)

Ophiuroidea
Amphipholis squamata (della Chiaje, 1828)
Amphiura angularis (Lyman, 1879)
A. magellanica (Ljungman, 1866)
Homalophiura inornata (Lyman, 1878)
Ophiacantha pentagona Koehler, 1922
O. sollicita Koehler, 1922
Ophiactis hirta Lyman, 1882
Ophiomitrella conferta (Koehler, 1922)
Ophiomyxa sp. McKnight, 1984
Ophiopyren regularis Koehler, 1901
Ophiuroglypha irrorata (Lyman, 1878)

Holothuroidea
Ocnus calcarea (Dendy, 1896)
Pseudocnus laevigatus (Verrill, 1876)
P. leoninoides (Mortensen, 1925)
Pseudopsolus macquariensis (Dendy, 1896)

Crustacea
Copepoda
Tigriopus angulatus Brehm, 1935 (= *T. californicus* Baker, 1912) (3)

Isopoda
Exosphaeroma gigas (Leach, 1818) (2)
?Munna maculata (Leach, 1818) (2)
Pancoloides litoralis (van Höffen, 1914)
  (2)

Amphipoda
Allorchestes novizealandiae Dana, 1852 (2)
Hyale hirtipalma (Dana, 1852) (1)

Decapoda
Halicarcinus platanus (Fabricius, 1775) (2)
Lithodes murrayi Henderson, 1888 (2)

Annelida
Polychaeta
Abarenicola assimilis (Ashworth,1902) (2)
Platynereis magalhaensis Kinberg, 1866 (2)
Spirorbis aggregatus Caullery et Mesnil,
  1897 (2)

Boccardia polybranchia (Haswell, 1885)
  (1)
Cirratulus cirratus (O. F. Müller) (1)

Oligochaeta
Lumbricillus macquariensis Benham, 1905
  (4)
L. werthi (Michaelson, 1905) (4)
Marionina antipodum (Benham, 1905) (1)

Bryozoa
Barentsia aggregata (1)

Legend
(1) addition from Kenny and
    Haysom (1962)
(2) addition from Simpson
    (1976a)
(3) addition from Bennett (1971)
(4) addition from Jamieson
    (1968)
*interim identifications

APPENDIX 9

Nearshore fishes of Macquarie Island

Data from Williams (1988)

Benthic species
Family Squalidae
Somniosus microcephalus Bloch et Schneider 1801. (Worldwide distribution. One record from Macquarie Island.)

Family Muraenolepididae
Muraenolepis marmoratus Günther 1880

Family Macrorhamphosidae
Centriscops cf. *humerosus* (Richardson 1848)

Family Congiopodidae
Zanchlorrhynchus spinifer (Günther 1880). (One of the commonest fish near Macquarie Island. Most common in water <100 m deep and often seen in kelp zone. Known from all subantarctic islands)

Family Psychrotulidae
Ebinania macquariensis Nelson 1982. (Endemic to Macquarie Island)
'*Neophrynichthys*' *magnicirrus* Nelson 1977. (Endemic to Macquarie Island)

Family Nototheniidae
Dissostichus eleginoides Smitt 1898. (Exploited commercially. Common off Iles Kerguelen, Patagonia, Antarctic Peninsula and most subantarctic islands)
Notothenia rossii rossii Richardson 1844. (Juveniles live in zone limited by outer zone of giant kelp)

Notothenia squamifrons Günther 1880. (Widespread and common around all subantarctic islands. Fished commercially at South Georgia and Iles Kerguelen)
Paranotothenia magellanica (Forster 1801). (Commonest inshore fish at Macquarie Island. Small specimens major diet of king shag. Selectively eaten by gentoo penguins. Taken by king and royal penguins. Very widespread in subantarctic)
Paranotothenia microlepidota (Hutton 1875). (Typical of New Zealand region)

Family Harpagiferidae
Harpagifer georgianus georgianus Nybelin 1947. (Common in intertidal and sublittoral zone)

Pelagic species
Family Bathylagidae
Bathylagus sp.

Family Gonostomatidae
Cyclothone sp.
Maurolicus mulleri (Gmelin 1788)

Family Stomiatidae
Stomias gracilis Garman 1899

Family Myctophidae
Electrona antartica (Günther 1878)
Electrona carlsbergi (Tåning 1932). (Major part of diet of gentoo and king penguins.)

Electrona subaspera (Günther 1864). (One
of commonest pelagic species near
Macquarie Island. Major component of
summer diet of gentoo penguins)

Krefftichthys anderssoni (Lönnberg 1905).
(Major dietary component of
Macquarie Island penguins)

Gymnoscopelus (*Gymnoscopelus*) *bolini*
Andriashev 1962

Gymnoscopelus (*Gymnoscopelus*) *braueri*
(Lönnberg) 1905

Gymnoscopelus (*Gymnoscopelus*) *nicholsi*
(Gilbert) 1911

Gymnoscopelus (*Nasolychnus*)
hintonoides Hulley 1981

Gymnoscopelus (*Nasolychnus*)
microlampas Hulley 1981

Gymnoscopelus (*Nasolychnus*) *piabilis*
(Whitley 1931). (One *Gymnoscolus* sp.
is a significant component of diet of
gentoo penguins)

Protomyctophum (*Protomyctophum*)
bolini (Fraser-Brunner 1949)

Protomyctophum (*Hierops*) *parallelum*
(Lönnberg 1905)

Family Scopelosauridae
Scopelosaurus hamiltoni (Waite 1916)

Family Paralepididae
Magnisudis prionosa (Rofen 1963)

Family Alepisauridae
Alepisaurus brevirostris Gibbs 1960

Family Bramidae
Brama sp.

Family Gempylidae
Paradiplospinus gracilis (Brauer 1906)

APPENDIX 10

Birds of Macquarie Island

Prepared from Anon. (1987). *Macquarie Island Nature Reserve. Visitor's Handbook*

Sphenisciformes
Spheniscidae
Aptenodytes patagonicus Miller, 1778.
  King penguin (abundant, breeds)
Eudyptes chrysocome (Forster, 1781).
  Rockhopper penguin (abundant,
  breeds)
Eudyptes robustus Oliver, 1953. Snares
  crested penguin (rare vagrant)
Eudyptes schlegeli Finsch, 1876. Royal
  penguin (endemic, abundant, breeds)
Eudyptes sclateri Buller, 1888.
  Erect-crested penguin (occasional
  vagrant)
Pygoscelis adeliae (Hombron et Jacquinot,
  1841). Adelie penguin (very rare
  vagrant)
Pygoscelis antarctica (Forster, 1781).
  Chinstrap penguin (rare vagrant)
Pygoscelis papua papua (Forster, 1781).
  Gentoo penguin (abundant, breeds)

Procellariiformes
Diomedeidae
Diomedea chrysostoma Forster, 1785.
  Grey-headed albatross (few, breeds)
Diomedea exulans L., 1758. Wandering
  albatross (rare, breeds)
Diomedea melanophrys Temminck, 1828.
  Black-browed albatross (few, breeds)
Phoebetria fusca (Hilsenberg, 1822). Sooty
  albatross (very rare vagrant)
Phoebetria palpebrata Forster, 1785.
  Light-mantled albatross (common,
  breeds)

Procellariidae
Daption capense australe (Mathews, 1913).
  Cape petrel (occasional vagrant)
Fulmarus glacialoides (Smith, 1840).
  Antarctic fulmar (very rare vagrant)
Halobeana caerulea (Gmelin, 1789). Blue
  petrel (common, breeds)
Macronectes giganteus (Gmelin, 1789).
  Southern giant petrel (common,
  breeds)
Macronectes halli Mathews, 1912.
  Northern giant petrel (common,
  breeds)
Pachyptila belcheri (Mathews, 1912).
  Thin-billed prion (rare, may breed)
Pachyptila desolata macquariensis
  (Mathews, 1812). Antarctic prion
  (abundant, breeds)
Pachyptila turtur (Kuhl, 1820). Fairy prion
  (few, breeds)
Pelecanoides georgicus Murphy et Harper,
  1916. South Georgia diving petrel (two
  alleged specimens in Tring Museum,
  collected 1899)
Procellaria cinerea Gmelin, 1789. Grey
  petrel (rare, may breed)
Pterodroma inexpectata (Forster, 1844).
  Mottled petrel (very rare vagrant)
Pterodroma lessonii (Garnot, 1826).
  White-headed petrel (abundant,
  breeds)
Pterodroma mollis (Gould, 1841).
  Soft-plumaged petrel (rare, may breed)
Puffinus assimilis Gould, 1838. Little
  shearwater (very rare vagrant)
Puffinus griseus (Gmelin, 1789). Sooty
  shearwater (common, breeds)

Puffinus tenuirostris (Temminck, 1835).
Short-tailed shearwater (very rare
vagrant)
Thalassoica antarctica (Gmelin, 1789).
Antarctic petrel (very rare vagrant)

Oceanitidae
Garrodia nereis (Gould, 1840).
Grey-backed storm petrel (rare, may
breed)

Pelecanoididae
Pelecanoides urinatrix (Gmelin, 1789).
Common diving petrel (rare, breeds)

Pelecaniformes
Phalacrocoracidae
Phalacrocorax albiventer purpurascens
(Brant, 1837). Blue-eyed cormorant
Phalacrocorax carbo novaehollandiae
Stephens, 1826. Black cormorant (rare,
vagrant) (endemic, common, breeds)

Sulidae
Morus serrator (Gray, 1843). Australasian
gannet (very rare vagrant)

Ardeiformes
Ardeidae
Ardea novae-hollandiae Latham, 1790.
White-faced heron (rare vagrant)
Ardeola ibis (L., 1758). Cattle egret (very
rare vagrant)
Egretta alba modesta (J. E. Gray, 1831).
Large egret (very rare vagrant)
Egretta garzetta (L., 1766). Little egret
(very rare vagrant)

Anseriformes
Anatidae
Anas gibberifrons gracilis Buller, 1869.
Grey teal (common vagrant during
drought in Australia)
Anas platyrhynchos platyrhynchos
L., 1758. Mallard (few, breeds and
hybridises with black duck)
Anas superciliosa superciliosa Gmelin,
1789. Black duck (common, breeds)

Accipitriformes
Circinae
Cirus aeruginosus gouldi Bonaparte, 1850.
Swamp harrier (very rare vagrant)

Gruiformes
Rallidae
Fulica atra australis Gould, 1845.
Australian coot (rare vagrant)
Gallirallus australis scotti (Ogilvie-Grant,
1905). Stewart Island weka
(introduced, few, breeds)
Porzana pusilla (Pallas, 1776). Marsh
crake (very rare vagrant)
Rallus philippensis macquariensis Hutton,
1879. Pacific banded rail (extinct
endemic)

Charadriformes
Charadriidae
Pluvialis squatarola (L., 1758). Grey
plover (very rare vagrant)

Arenariidae
Arenaria interpres (L., 1758). Ruddy
turnstone (very rare vagrant)

Scolopacidae
Calidris canutus (L., 1758). Knot (very
rare vagrant)
Calidris ruficollis (Pallas, 1776).
Red-necked stint (very rare vagrant)
Gallinago hardwickii (Gray, 1831).
Latham's snipe (very rare vagrant)
Limosa lapponica (L., 1758). Bar-tailed
godwit (very rare vagrant)
Tringa nebularia (Gunnerus, 1767).
Greenshank (very rare vagrant)

Recurvirostridae
Himantopus himantopus (L., 1758). Pied
stilt (rare vagrant)

Phalaropodidae
Phalaropus lobatus (L., 1758). Northern
phalarope (very rare vagrant)

Stercorariidae
Stercorarius skua lonnbergi (Mathews,
1912). Great skua (common, breeds)

Laridae
Larus dominicanus Lichtenstein, 1823.
  Kelp gull (few, breeds)

Sterninae
Sterna paradisaea Pontoppidan, 1763.
  Arctic tern (very rare vagrant)
Sterna vittata Gmelin, 1789. Antarctic tern
  (few, breeds)

Psittaciformes
Psittacidae
Cyanorhamphus novaezelandiae erythrotis
  (Wagler, 1832). Macquarie Island
  parakeet (extinct endemic)

Apodiformes
Apodidae
Apus pacificus pacificus (Latham, 1801).
  Fork-tailed swift (very rare vagrant)
Hirundapis caudacutus caudacutus
  (Latham, 1801). Spine-tailed swift (rare
  vagrant)

Passeriformes
Hirundinidae
Hirundo neoxena Gould, 1852. Welcome
  swallow (very rare vagrant)

Turdidae
Turdus merula L., 1758. Blackbird (rare
  vagrant)
Turdus philomelos Brehm, 1831. Song
  thrush (very rare vagrant)

Zosteropidae
Zosterops lateralis (Latham, 1801).
  Silvereye (very rare vagrant)

Fringillidae
Acanthis flammea (L., 1758). Redpoll
  (common, breeds)
Carduelis carduelis britannica (Hartert,
  1903). Goldfinch (very rare vagrant)

Sturnidae
Sturnus vulgaris L., 1758. Common
  starling (common, breeds)

APPENDIX 11

Marine mammals of Macquarie Island

Prepared from Anon. (1987). *Macquarie Island Nature Reserve Visitor's Handbook*

Cetacea

Balaena glacialis Borowski 1781. Southern right whale (rare sightings)

Globicephala melaena Traill 1809. Longfin pilot whale (sightings and strandings)

Hyperodon planifrons Flower 1882. Southern bottlenose whale (one stranding on west coast)

Orcinus orca (L. 1758). Killer whale (regular sightings, occasional strandings)

Physeter macrocephalus L. 1758. Sperm whale (rare sightings)

Ziphius cavirostris Cuvier 1823. Cuvier's beaked whale (one skull known from west coast)

Pinnipedia

Arctocephalus forsteri (Lesson 1828). New Zealand fur seal (annual visitor. Not known to breed)

Arctocephalus gazella (Peters 1875). Antarctic fur seal (cows only, breeding with *A. tropicalis*)

Arctocephalus tropicalis (Gray 1872). Subantarctic fur seal (small breeding group)

Hydrurga leptonyx (Blainville 1820). Leopard seal (regular visitor, mainly in winter)

Leptonychotes weddelli (Lesson 1826). Weddell seal (irregular visitor)

Lobodon carcinophagus (Hombron et Jacquinot 1842). Crabeater seal (irregular visitor)

Mirounga leonina (L. 1758). Southern elephant seal (large breeding population)

Phocarctos hookeri (Gray 1844). Hooker's sea lion (winter visitor)

APPENDIX 12

Introduced mammals established on Macquarie Island

Prepared from Anon. (1987). *Macquarie Island Nature Reserve Visitor's Handbook*

Lagomorpha
Oryctolagus cuniculus (L. 1758).
    European rabbit

Carnivora
Felis catus L. 1758. Feral cat

Rodentia
Mus musculus L. 1758. House mouse
Rattus rattus (L. 1758). Black rat

Index of genus and species names

This index includes only genera and species mentioned in the text. It does not include taxa that are listed in the appendixes but not mentioned in the text.

Subject index